ARCHIPEL
INDONESIA
KINGDOMS
OF THE SEA

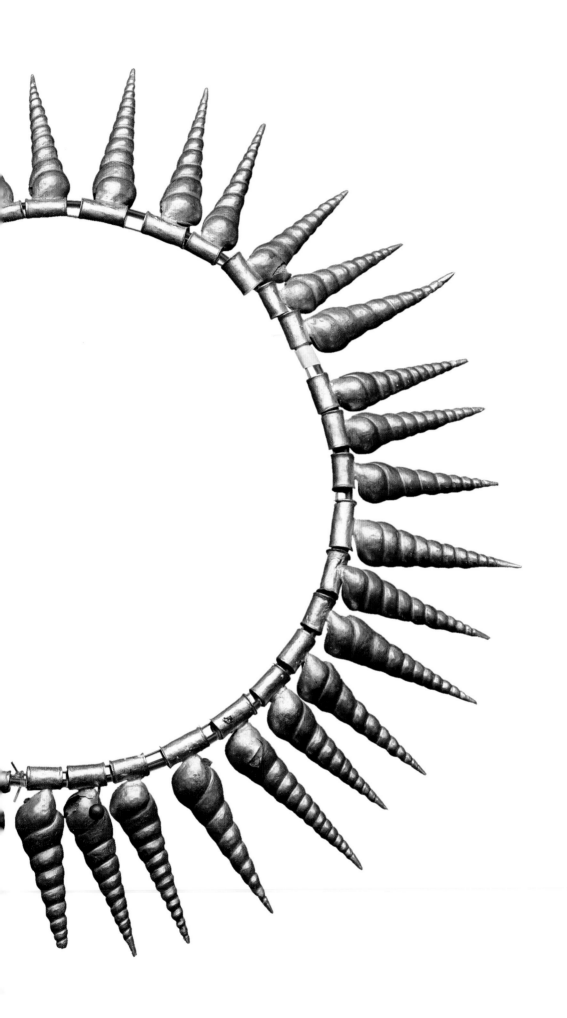

snoeck

ARCHIPEL

INDONESIA
KINGDOMS
OF THE SEA

EUROPALIA
ARTS FESTIVAL
INDONESIA

TABLE OF CONTENTS

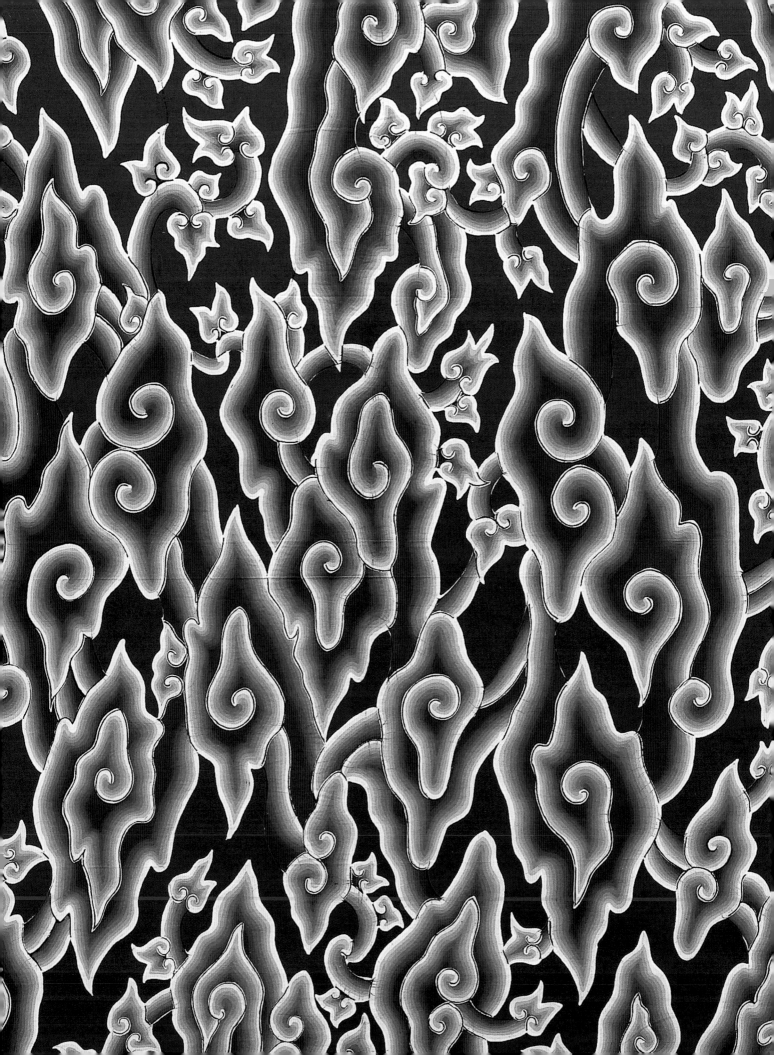

Assalamu'alaikum wa rahmatullaahi wa barakatuh
Peace and prosperity to all

A cultural exhibition, whether of material or immaterial content, is a strategic form of cultural diplomacy between communities and countries. The purpose of cultural diplomacy is not only to showcase the highest attainment of a community's cultural maturity, but also to promote a better understanding of it developed.

In the framework of economic and cultural globalisation, local cultures are of major interest to the people of the modern world, shackled by global – by which is meant Western – culture. Consequently, the preservation and development of a nation's heritage should be understood as the protection of priceless assets that can serve as tourist attractions and a reference for new generations. Importantly, in addition to encouraging culture-based economic development, heritage preservation also strengthens the national identity.

The Archipel exhibition, with its emphasis on Indonesia's maritime culture, offers strategic benefits. The objects displayed are not merely artefacts of artistic and cultural value, they are arranged within the context of Indonesia's maritime history. Visitors are made aware that Indonesia was once a great sea-going nation, a major aspect in the history and identity of this nation.

Geographically, Indonesia was destined to be a maritime nation. Reflected in the historical trajectory of the country, this fact is also echoed in the distribution of the population, a great proportion of which inhabits coastal zones. All of Indonesia's cultural heritage can be traced back to its origin as a nation that traded with its international partners by sea. Pieces of pottery, vestiges of temples, and statues that would seem to be the products of an agrarian nation are also the heritage of shipping and trading activities, and recognition of this fact will contribute to the development of our potential as a great maritime nation of the future. It is time to widen our gaze from our current focus on contemporary urban and terrestrial culture to accept our maritime roots.

Archipel. Kingdoms of the Sea is the largest event staged by the National Museum of Indonesia to celebrate the country's maritime culture. The cooperation between the National Museum of Indonesia and La Beverie Museum in Liège, Belgium, the Cultural Heritage Preservation Office of Jambi Province, 'Si Ginjei' Museum of Jambi Province, 'Ruwa Jurai' Museum of Lampung Province, and the Cultural Heritage Preservation Office of South Sulawesi Province has resulted in a great achievement. It is my hope that Archipel will demonstrate that the vision of Indonesia's historical role as one of the world's maritime axes is based on fact, and in accordance with the Indonesian people's identity of their nation.

Wassalamu'alaikum wa rahmatullaahi wa barakatuh

HILMAR FARID
Director General of Culture Ministry
of Education and Culture
Republic of Indonesia

An area of confluence and encounters, the Indonesian archipelago has always been one of the most important crossroads of world trade. Here Austronesian ships, Arabian Dhows, Chines junks, Iberian caravels and vessels of the East India Company once dropped anchor, long before the container ships and oil tankers of today.

Europalia Indonesia is therefore very proud to present this exposition showing the all-important role of the sea for the country. It is the fruit of an intense collaboration between Indonesian, Belgian, French and Dutch institutions and scientists, La Boverie and the city of Liege.

Archipel. Kingdoms of the Sea shows exceptional national treasures of which many are making their first ever journey outside Indonesia.

In the name of the whole Europalia team, we wish you an excellent visit full of enchanting discoveries.

KOEN CLEMENT
General Manager Europalia

The concept for the *Europalia Indonesia Festival* has been built around three themes that run throughout the whole of the programme: ancestors and rituals, trading, and biodiversity. As a part of this programme, The City of Liège is delighted for the opportunity to stage the major exhibition: *Archipel. Kingdoms of the Sea*.

Located at the heart of Asia – 'south of China, and east of India' – Indonesia and its 17,000 islands extend over a space equivalent to that of the European Union. At the Pacific intersection of the China Sea and the Indian Ocean, a region of meetings and convergence, the Indonesian archipelago has been one of the most important crossroads for global trade, where Austronesian boats, Arab dhows, Chinese junks, Iberian ships, and other vessels owned by the East India companies all took berth, long before the container ships and petrol tankers of today.

The history of this archipelago is that of a host of links and connections, where what is near and what is distant rub shoulders, and are set in competition by the ever-present dynamics of the maritime world. The sea no longer seems like something that separates or divides, and the monsoon winds have made this crossroads an obligatory place to stop for the merchants, artisans, and priests who have all left their mark on the myths, monuments, arts, and traditions of contemporary Indonesia.

Superimposed, interwoven, and reinterpreted by a succession of rich and complex societies, these external contributions forged the multiple worlds that this exhibition invites us to discover, with trade and the sea acting as the common threads and an exceptional collection of major works as the markers of a history to be shared and admired.

We are very grateful to our key partner, the National Museum of Indonesia in Jakarta, and its generous, energetic teams for helping us to discover so many exceptional works and for allowing us to show them at La Boverie. We should also like to thank the Royal Museum of Mariemont, the Research and Information Group on Peace and Security, for being helpful and friendly partners in this adventure, as well as the French School of the Far East and its active representative, Pierre-Yves Manguin, for bringing us the insight needed to gain an understanding of events and phenomena.

The Maritime Museum Rotterdam, the Paris Maritime Museum, the University of Utrecht Library, the EFEO, and a number of private collectors have all been very supportive in lending important works that contribute to the exceptional wealth and variety of this exhibition.

We are convinced that the exhibition *Archipel. Kingdoms of the Sea* will help to bring a new vision of Indonesia to Belgium and Europe, that of an archipelagic state with a major role to play in the world of tomorrow.

WILLY DEMEYER
Mayor of Liège

JEAN PIERRE HUPKENS
Alderman for Culture and Urban Planning,
City of Liège

INTRODUCTION

Nusantara is an ancient term for the archipelago that makes up most of what is now the Republic of Indonesia. Indonesia is located in a strategic position, between the tropical waters of the Indian Ocean and the Pacific, between South East Asia and the north of Australia. This position has allowed Indonesia to become the crossroads and the historical point of entry between the West and the East. Nusantara's role as a gateway and catalyst for maritime trade between the two giants of India and China continues to grow. The Indonesian people have proven they are capable of becoming an active participant in economic and maritime interactions with India, the Middle East, the South East of Asia, and China. This relationship has made it possible for Indonesian culture to exert an influence on the Indian Ocean, a process entailing a cultural enrichment that is both tangible and non-tangible, in terms of beliefs, language, political systems, architecture and art, etc.

The importance of Nusantara's strategic role has continued to grow since the onset of economic and cultural globalisation linked to interactions between the Islamic and the Christian worlds. The economic and cultural exchanges conducted via the straits of Malacca and Sunda involve the Middle East, India, China and Europe, as much as the multitude of ethnic groups that make up the archipelago. Currently, more than 94,000 boats transport around one quarter of the world's maritime trade through these straits every year.

As well as occupying a strategic position, the Republic of Indonesia is also the largest archipelagic region in the world. The distance separating the extremities of the Indonesian archipelago is greater than the distance between London and Moscow or that between New York and San Francisco. This particular situation allows Indonesia to present itself as a maritime nation unique in the world. Around 80% of the surface area of the Republic of Indonesia is covered by the sea, with the rest forming a string of 17,000 islands. The land mass represents 1.92 million km² and the maritime territory a further 3.1 million km². The maritime part included in the EEZ (Exclusive Economic Zone) forms an area of 2.7 million km². Moreover, Indonesia possesses the longest coastline in the world, measuring 81,000 km.

The vast territory just described is inhabited by hundreds of ethnic groups using different types of languages and dialects. The main ethnic group is that of the Javanese, who form 40.2% of the Indonesian population. This group is located particularly in Central Java, East Java, Yogyakarta, Bengkulu, Lampung, Jambi, North Sumatra, Riau, South Sumatra, and in Borneo. The Sundanese inhabitants amount to some 15.5% of the population, and are mainly spread around West Java and Banten. The Batak people represent almost 3.6% of Indonesians, and they mainly populate the region of North Sumatra. The people from the south of Sulawesi account for 3.2%, the Madurese some 3.03%, and the Minang 2.7%, etc.

Indonesians recognize the complexity of their ethnic diversity and the need for tolerance, coexistence, harmony and peace are common values. This awareness of diversity was clearly expressed by the Javanese poet Mpu Tantular in his book Sutasoma, which was written in the fourteenth century. The author lived during the golden age of the Majapahit Empire, whose territories are believed to be as great or even greater in size than present-day Indonesia, covering Brunei, Singapore, Malaysia, and part of the Philippines. In his book, Mpu Tantular unveils the slogan 'Bhinneka Tunggal Ika', which means Unity in Diversity. This slogan draws on his own thinking and reflects the spirit of his time. The Majapahit Empire was then one of the largest territories in Indonesia, comprising a multitude of ethnic territories with diverse cultures, religions and economic contexts. As a consequence, the priority was to create a life of harmony amidst the diversity. Mpu Tantular himself practiced Tantric Buddhism and led a peaceful life in the kingdom of Majapahit, despite its Hindu influence. 'Bhinneka Tungga Ika' remains the slogan of the Republic of Indonesia even now, a veritable shield against internal problems.

From this sense of awareness of evolving in an archipelagic world comes the idea of Nusantara, a concept referring to the union of islands by the sea that has been evolving since pre-modern times. This concept is already present in the book Negarakertagama, which was written by Mpu Prapanca at the height of Majapahit in the fourteenth century. It refers to a political idea, namely that Nusantara and Indonesia are as one. The concept was rediscovered by nationalist Indonesian leaders who will go on to make it the heart of the Indonesian spirit. Nusantara is still recognized as one of the other names used to designate Indonesia.

The spirit of the archipelago was very clearly defended following the proclamation of Indonesian independence of 17 August 1945, at a time when the Dutch were still trying to retain control over West Papua. The spirit of Nusantara led the fight against colonialism and became the cornerstone in the construction of the Indonesian state as a maritime country of prime importance. With the Djuanda Declaration of 13 December 1957, the Indonesian government declared that the principle of an archipelagic state had been applied, as recognized by the United Nations. This declaration established the territory of Indonesia covering an area of territorial waters extending 12 nautical miles beyond a line connecting the extremes of the most distant islands.

In 1973, the Indonesian government declared that at sea, the Indonesian continental shelf would remain a part of the Indonesian region to a depth of 200 meters. In addition, on 21 March 1980, the Indonesian government officially announced the implementation of the Declaration of the EEZ, an area extending 200 miles from the Indonesian shoreline. This policy was followed by the Caribbean, Peru, Brazil, El Salvador, Chile, Ecuador, the United States, Canada, Norway, Mexico, and India. The ruling was finally adopted by the conference of the United Nations as Law of the Sea II on 30 April 1982.

The Indonesian government formulated the idea of an Indonesian nation and a state called Wawasan Nusantara ('archipelagic concept') – a state based on the formulation of a strategy. It is the history of Indonesia – equally full of dynamism in times of peace as it is in times of crisis – that inspired its construction as a maritime nation. For Indonesia, a maritime nation is a nation whose power relies as much on transport and trade as it does on defensive and security forces and state-of-the-art naval technology. It must be able to exploit its terrestrial and maritime potential in the context of geopolitical dynamics.

THE ANCIENT PERIOD

3000 BCE – EARLY CENTURIES

THE STATE OF THE DIASPORA

The archipelago was formed ten million years ago. Fossils from the basin of the Bengawan Solo river in Java attest to the existence of hominids 1.8 million years ago. It was in 1891 that Eugène Dubois made his discovery in Trini (East Java) of what will later be known as Java Man. This was *Homo erectus* (*Pithecanthropus erectus*), the species that probably gave birth to us, *Homo sapiens*. *Homo erectus* had in all likelihood migrated from Africa to different parts of the world.

Java Man would be about 100,000 years old, which locates him in a period of transition between *Homo erectus* and *Homo sapiens*, who will make his appearance some 60,000 years later. It is possible to imagine these people forming an Australo-Melanesean ethnic group, from which the ancestors of the Melanesians of Papua New Guinea, the Australian Aborigines, and the Negritos of the Malay Peninsula and the Philippines would all originate.

The first humans were subject to 40,000 years of changes in climate, which encouraged migration, especially among the Papua-Melanesians. Because of the rising sea levels that followed the end of the last glaciation, humans regrouped in the drier areas of the eastern archipelago. 7,000 years before our era, the Papua-Melanesians developed a form of agriculture, based on taro, bananas and tubers.

Originally from southern China (Taiwan), the Austronesians arrived in the archipelago around 4,000 years BCE. It is supposed that they moved to the archipelago and the Pacific in a series of successive waves. According to Heine Geldem, the Austronesians were familiar with the construction of outrigger boats, and knew how to sail, make pottery, polish tools like square axes, build houses, grow rice, raise animals, make a form of cloth by pounding bark, build megalithic structures, and develop forms of art. Together these cultures constitute *the Culture of the Square Axe* and our understanding of their history is based on the distribution of stone axes and picks with a sharpened surface.

The Taiwanese origin of the Austronesians has been confirmed by findings connected with the Lapita civilization in Western Melanesia, and particularly in the Bismarck archipelago, which is characterized by a red, generally polished form of pottery decorated with a comb-like tool that stamps designs into the wet clay. Experts propose that the Lapita culture was brought in by Austronesians arriving from Asia. The discovery of the Kalumpang pottery (in South Sulawesi), very similar to that of Lapita, supports the hypothesis of movement and trade taking place by sea across distances of thousands of kilometres. These long-distance movements could not take place without a knowledge of astronomy, the seasons, and the prevailing winds and currents. The acquisition of these capacities by the populations of the archipelago enabled them to become maritime communities and to interact very early with other cultures from South East Asia. Several centuries before the beginning of the Common Era, the inhabitants of Nusantara mastered the art of metalwork, as is evidenced by the influence of Dong Son culture in northern Vietnam. There are nekara, moko, axes, caskets, and bronze jewellery that date back to this culture from the age of metals.

13

The waves of the Austronesian diaspora separated into two different tides as they reached the south of Sulawesi. One tide flowed towards Kalimantan, the Malay Peninsula and Champa (in the south of Vietnam). Another tide headed towards Papua, where it in turn split in two: one migration continued eastwards to Fiji and Tonga (1500 BC). The descendants of these people now inhabit the Pacific Islands as far away as New Zealand, Easter Island and Hawaii. Another migration headed West and reached Nusa Tenggara, Java and Sumatra.

In the period that followed, between the fifth and twelfth centuries CE, another diaspora spread to Madagascar as the result of commercial activities in the time of the kingdom of Srivijaya.

Within a few centuries, the Austronesians thus spread out over more than half the circumference of the Earth. Robert Cribb describes it as one of the most important human diasporas of all time.

Information on the relations that existed between the Melanesian and the Austronesian peoples is incomplete, although the physical appearance of modern Indonesians suggests a genetic mixing. The Austronesian populations did not replace the indigenous peoples: mixed alliances probably took place. On the other hand, ethnic groups may not have mixed with the Austronesians, even if they adopted the culture.

In the Indonesia of today, dozens of major ethnic groups and hundreds of smaller groups co-exist. Physical appearance, language, religion, names, place of birth and social traditions make it possible to distinguish between them. Ethnic classification can be more readily applied, however, to the descendants of recent immigrants (Chinese, Arabs, Europeans).

THE ROOTS OF DIVERSITY

The Austronesians share the same genealogical ancestry, and even at the time of their diaspora to Indonesia, they would have spoken the same language. The manner in which they ended up living scattered and separated lives resulted in the diversification of their culture and language. In Indonesia today, there are 200 Austronesian and 150 Melanesian (Papuan) languages. The similarity of some of the languages makes it possible to put forward hypotheses about the close history of

certain ethnic groups who probably had common ancestors. For example, the close ties between the Madurese – inhabitants of the island of Madura and neighbouring areas of East Java – and the Malay language make it possible to say that the island was closer to the migrants of Sumatra than to those of Java. The survival of Papuan languages in Timor and its neighbouring islands confirms the archaeological evidence that this area was an important centre for the Melanesians before the arrival of the Austronesians. Political conditions have also influenced linguistic diversity in the history of Indonesia. Before the expansion of the kingdom of Srivijaya, a great number of languages were spoken in the north and south of Sumatra, as well as in the Malay Peninsula. However, from the seventh to the eleventh century, the cultural influence of Srivijaya proved decisive in reducing that number. In Aceh, on the other hand, Austronesian linguistic elements have remained well preserved.

The succession of little kingdoms on the island of Borneo have never united into any form of powerful empire. The various ethnic groups there speak different languages. However, this does not mean that they are isolated societies: the communities in the north of Borneo have maintained regular contact with other descendants of the Austronesians of Champa, between Vietnam and China. Similarly, the Ma'anyan language from the southeast of Kalimantan has a link with the languages of Madagascar. The lack of political integration in Papua has also contributed to the ethnic and linguistic diversity of the island, even between groups that are geographically close.

From the fifteenth century on, Malay became the dominant language in the coastal zones of Nusantara, in part driven by Muslim merchants coming from the Middle East. Malay was the language of the diffusion of Islam in Indonesia.

It was the Javanese language that dominated the densely populated areas of Central Java and East Java where a series of agricultural kingdoms had developed since the eighth century. Javanese is a complex language, both in terms of its grammar and its social use: the vocabulary of a Javanese speaker depends on his own social status, the social status of his interlocutor and the links that they share between them: a discussion is therefore impossible until the social positions of all parties are understood.

Javanese is too complex for commercial use, so Malay established itself as the language of trade and the spread of Islam. However, Javanese is used as a language of instruction in Islamic schools (*pesantren*).

TOWARDS THE ERA OF INTERNATIONAL COMMERCE

The maritime experience of the Austronesian ancestors is a precious socio-cultural resource that has played an active role in the trade between India and China, through the Straits of Malacca, from the onset of the Common Era. Four civilizations began to emerge from the fifth century BCE: Rome, Egypt, Mesopotamia, China, and India/Persia. Two centuries later, trade between these civilizations is still sporadic. Intercontinental trade began to develop from the second century BCE onwards, linking China and the Mediterranean via Central Asia. Silk, gold and silver, textiles, metals and other luxury goods are beginning to be exchanged: this is the beginning of the Silk Road.

Though the trade routes were put in place a few thousand years ago, they were initially limited to relations between India and Egypt via the Persian Gulf and the Red Sea. Trade mainly consists of spices (pepper from Malabar and cinnamon from Sri Lanka), along with textiles and precious metals. The road from India to Indonesia and China does not yet exist.

The extension of the Spice Route to the East began at the start of the Common Era. The gold trade is then disrupted. India, which had previously imported large quantities of gold from Rome, is confronted with a ban on exports of precious metals imposed by the Emperor Vespasian (69–79 CE). The Indians then go off in search of new sources of gold to the East, towards Indonesia, no doubt influenced by the literature suggesting the existence of an Eldorado (*Suvamna Dvipa*). The Indians soon discovered that the spices of the archipelago were of better quality and less expensive than the ones they knew. They also discovered sandalwood, camphor, incense, and precious woods. The Indians traded these precious commodities for the fabrics they produced at low cost. From this point on, the Indonesians became fully involved in international trade, where their qualities as sailors enjoyed full rein.

⌄ 2. Dish

Paleometallic period
Buni, Bekasi, West Java
Earthenware, 27.5 x 5.5 cm
MNI 7049

Four thousand years ago, Austronesian migrants introduced a rudimentary technique into the archipelago to make pottery. Based on the use of a paddle and an anvil, this development marked the start of a new era. Use of the potter's wheel came at a later date. This dish, made on a wheel, originated in Arikamedu, a coastal port in southern India. It has all the characteristics of the production of this city: its form, the surface polished to shininess, its brownish-red colour with a grey interior, and a circular décor typical of flat dishes in the Roman-Indian tradition. Dating from the 2nd or 3rd century, dishes of this type were used to hold food or burial provisions, and could also be used as a means of exchange.

> 3. Pottery

Kei Islands, South-East Maluku
33 x 29 cm
MNI 14283

This container was made using the 'paddle and anvil' technique and has been painted with geometric motifs using natural pigments. This type of vessel was first made exclusively on the Banda Islands but, after 1621, refugees from Banda began production on the Aru and Kei islands. Only women were allowed to make these containers as they were a symbol of fertility and an image of the uterus. They were used for both daily and ceremonial purposes: utilised to conserve foodstuffs, they were also supposed to resist the powers of human spirits.

∧ 1. Axe (*belincung*)

Neolithic
Ciamis, West Java
Chalcedony, 3 x 17 x 6.3 cm
MNI 4390

This *belicung* axe is a variant of the *beliung* axes dating from the Neolithic period: the two types are rectangular but the former is pointed. *Belincung* axes have been found everywhere in Indonesia and are an indicator of the Neolithic period. Cutting and polishing techniques were introduced by Austronesian migrants 4000 years ago. Used in agriculture when they were made from stone, these axes became ritual objects or a means of exchange when made from semi-precious materials like chalcedony.

> 4. Bead necklace

Paleometallic period
Besuki, East Java
Clay, glass, 4.6 x 22 cm
MNI 1316

The use of beads dates back to the era of prehistoric cave dwelling. At that time, jewellery was made using shells. Later, during the Paleometallic, beads were made from various materials (carnelian, glass, earthenware, etc.) and in different forms and colours.

< 5. Drum (*nekara)*

Paleometallic period
Sangeang, West Nusa Tenggara
Bronze, 92 x 117 cm / 157.5 kg
MNI 3364

Nekara are made in three parts: the shoulder, belly and feet. They are of two main types: either indigenous or made elsewhere. In 1902 the second type was subdivided by Austrian ethnologist Franz Heger into four categories in his book *Alte Metalltrommeln aus Südostasien*. This is the type Heger IV. Whereas the upper part is decorated with a central star, the walls of the drum feature birds, boats, elephants, horses, fish and humans. These motifs form a link between this drum, a major example of its type, and those from the Dong Son culture of Vietnam that developed from 2000 BCE and spread through Indonesia around 500 BCE. This *nekara* was probably brought back to Indonesia from a visit made for commercial purposes.
Called *Makalamau* by the inhabitants of Sangeang, these drums were considered to attract rain – which is why there are images of frogs on the upper rim – but could also be used as war drums, a means of exchange and even as funerary urns.

ᵛ 6. Bracelet

Paleometallic period
Bangkinang, Riau, Sumatra
Bronze, 2.1 x 8 cm
MNI 6140

As it introduced the working of metal, the mastery of fire by the first human beings represented a major step forward in the progress of civilisation. From the Stone Age, mankind entered the Metal Age. The metals worked were bronze, iron, gold and silver. In Indonesia the artefacts are above all made of bronze, though objects of iron have been found in smaller quantities. Metalworking was linked with the integration of products and knowhow from the Dong Son culture (see cat. 5): many types of object were imported but some artefacts, like this one, were made locally. This bracelet has a decoration formed of two lateral rows of conic protuberances on either side of a chevron motif. Symbols of prestige, these bracelets were sometimes worn by placed on the bodies of the dead when they were buried.

19

ᵛ 7. Barkcloth

Sentani, Papua province, New Guinea
Bark, 159 x 135 cm
MNI 24171

The use of bark for weaving, known since the Neolithic, spread right across the archipelago. Stones used to beat the bark have been found in Ampah (Kalimantan), Kalumpang, Minanga Sipakka, Langkoka, Poso (Sulawesi) and Papua-New Guinea. These stone beaters vary between 10 and 20 cm in length. The inner face is grooved lengthwise or in the form of a cross. This type of beater is still used in Kalimantan and Papua-New Guinea, but it is made from hard wood. Certain peoples in Kalimantan, Sulawesi and Sumatra continue to use barkcloth in ceremonies and for creating artworks. Papuan women still make cloth using the bark of *Ficus variegate*. They fell the trees and do all the necessary work with the exception of the decoration, which is the responsibility of men.

Barkcloth in the region around Lake Sentani is typified by an interlacing motif called *fouw*, which symbolises the links between the sky and the land, the ancestors and their descendants, the chiefs and the community, and the latter's cohesion. The execution of the motif reflects on the wealth of the clan or family. Among the colours used, red represents courage, white brotherhood, and black fertility and prosperity.

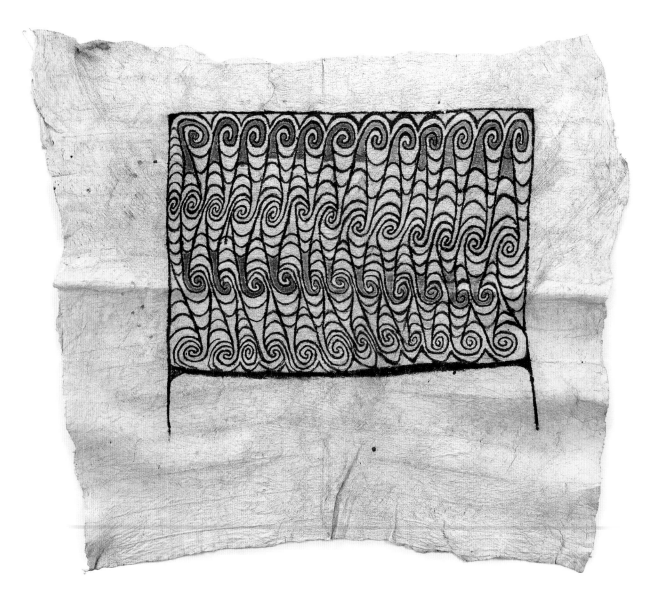

⌄ 8. Pickaxe

Neolithic
Menes, West Java
Chalcedony, 1.4 x 10.8 x 5.8 cm
MNI 2896

An Austronesian vestige (Utomo et al., 2016: 26), this pickaxe has never been used. It was in all probability an amulet or made for barter rather than a real tool.

⌄ 9. Necklace

Paleometallic period
Pala, West Sumba, East Nusa Tenggara
Clay, glass, 1 x 14 cm
MNI 2163

Austronesian (Utomo et al., 2016: 26), this necklace of round, oval and cylindrical beads decorated in different fashion was either an item of jewellery or a burial object, or both.

⌄ 10. Beads

Paleometallic period
Pasemah, South Sumatra
Carnelian, 5 x 2.5 cm
MNI 3214, 3215 and 3216

Beads like these, in truncated pyramidal form at either end, are often found in burial sites.

< **11. Bracelet**

Paleometallic period
Kalioso, Central Java
Carnelian, 1 x 8 cm
MNI 2147

Jewel and/or burial object.

< **12. Pot (*periuk*)**

Neolithic
Buni, Bekasi, West Java
Clay, 13 x 14.5 cm
MNI 6977

Made using the paddle and anvil technique, and decorated with an engraved trellis pattern, this pot was used to contain foodstuffs.

∨ **13. Container (*kendi*)**

Neolithic
Karawang, West Java
Clay, 15.5 x 20 cm
MNI 7005

With its unusual form, this container was either used for practical purposes (to hold water) or had a religious function.

⌃ 14. Bowl

Kei Islands, Maluku
Clay, 7 x 19.5 cm
MNI 14290

The Kei Islands in the Maluku were influenced by Austronesian culture. Making pottery was the responsibility of the women, and their production was one of the best in eastern Indonesia.

⌃ 15. *Kyokkenmodinger*

Mesolithic
Tanjung Morawa, North Sumatra
28.3 x 16.5 cm
MNI 921 (1)

These shells are a solidified mass of food remains. Also found in eastern Aceh province, mounds of these shells may attain a height of 5 metres (Wiradnyana, 1999: 1).

> 16. *Moko*

Paleometallic period
Flores, East Nusa Tenggara
Bronze, 58 x 36 cm
MNI 26531

The *moko* is a small indigenous drum. It is found in abundance in East Nusa Tenggara, where they are known by different names. The *moko* is often included in a dowry and is also used as a form of currency.

^ 17. Ceremonial axe

Paleometallic period
Rote Island, East Nusa Tenggara
Bronze, 89 x 51.3 cm
MNI 1441

Only three axes of this type have been found in Indonesia, of which one was lost in a fire in the Dutch Pavilion in the Paris Colonial Exhibition of 1931. The motif on the blade is of a human figure, a mask with magical powers.

^ 18. Ceremonial axe (*candrasa*)

Paleometallic period
Bandung, West Java
Bronze, 2.2 x 74.5 x 32.5 cm
MNI 1431

This *candrasa* is T-shaped. It has a short and long blade.

> 19. Ceremonial vase

Paleometallic period
Labuhan Moringan, Central Lampung
Bronze, 61 x 43 cm
Ruwa Jurai, Lampung Provincial Museum, 2440

In Indonesia, these prehistoric bronze vases have only been found on Java and Sumatra, and in very limited number. This one, which has remained intact, is in the form of a fish basket and is decorated with a variety of J-shaped forms.

˅ 20. Bracelet

Paleometallic period
Pagar Alam, Pasemah, South Sumatra
Bronze, 0.8 x 8 cm
MNI 2184

These small bronze bracelets were made either for exchange or for passing on to the next generation. The large versions were used for adornment or as burial goods, or both.

24

THE PRE-MODERN PERIOD
FIRST – FIFTEENTH CENTURIES

CONTEXT: FIRST CONTACTS WITH THE OUTSIDE WORLD

The archipelago's international trade grows rapidly. Indeed, the traditional tides of trade (especially between India and Siberia and between India and Rome) are subject to strong changes, leading to a reorientation of trade to the east, South East Asia in general, and to Nusantara in particular, which enjoys the reputation of being rich. Ancient Greeks called the region the Gold Country (*Khersonese d'Or*) and the Indians called it *Suvarnadvipa* (Gold Island). The spread of Indian religions in Indonesia (Hinduism and Buddhism) will also prove a determining factor in accelerating trade in the archipelago.

The spices (pepper, cinnamon, nutmeg, cloves) and medicinal plants were much appreciated by the Indians, even more so than the gold they had gone in search of. The Indian trade did not immediately penetrate through the Straits of Malacca but crossed the Kra Isthmus (now Thailand). The maritime trade continued, from the Gulf of Thailand to the Funan and further north to the Vietnamese region then under Chinese control. On this trade route, the Indians find silk fabrics. Trade led to the development of the Hindu kingdom of Funan at the southern tip of Vietnam, whose peak was between the second and fourth centuries.

In the fourth century of the Common Era, trade between the archipelago and China underwent an important development thanks to the presence of the ports on the east coast of Vietnam that made it possible to avoid Funan. The products of Nusantara (including wood, spices, and perfumes) were exchanged for the silk and porcelain that was being sold at the time in the archipelago and in the western countries.

New maritime routes were opened up through the straits of Malacca and Sunda, which were used not only by the sailors of Nusantara, but also by Indians and Westerners. The trading of goods was accompanied by a socio-political transformation: the Indonesians visited India in particular to study Hinduism and Buddhism.

Trade with China progressed more slowly: the Chinese were more keen to develop the already established land trade routes. It was the merchants of Nusantara who took the initiative for maritime trading with China, sending many ambassadors from the different kingdoms of the archipelago. It was only in the fifth century that China began to acquaint itself with what would become Indonesia. It was priests who went to India who first opened up these relations.

In 413, the Buddhist monk Fa Hsien embarked on an Indian ship bound for China. His narrative is recorded in *Fo-Kou-chi* (notes on the Buddhist countries). His ship was caught in a storm, forcing it to make a five-month stopover at *Yeh-p'o-t'i*, which may refer to Java. During the final journey, this priest writes of hearing passengers' astonishment at the delay in arriving to Canton.

This testimony allows us to deduce that direct journeys took place over long distances, and that the most seasoned travellers knew how long the journeys were supposed to take.

There are numerous references in Chinese palaces describing the commercial envoys of Nusantara. It is also worth mentioning that the sailors of the archipelago had also reached the west of the Malay peninsula, as evidenced by the journey taken by the Chinese Buddhist monk, I-Tsing, en route to India on a ship from Srivijaya in 671.

Before the arrival of Indian merchants to the ports of Nusantara, the inhabitants had already had the opportunity to meet them on the route linking India to Funan and China. They had seen how the Indian influence had enabled them to prosper economically and politically. This experience may have led the people of Nusantara to take the initiative in welcoming Indian beliefs in order to strengthen and reorder their society. This explains the emergence of many kingdoms, such as Kutai in the east of Kalimantan, Tarumanagara in Java West, Kalingga and Mataram in Central Java, and Malayu and Srivijaya in Sumatra.

These ties with India have not only enriched the political and religious domains, but also the arts of Nusantara. Some works from the archipelago have even proved more beautiful than the Indian originals.

INTERACTION AND TRADE WITH INDIA

Economic exchange
Spices became the driving force behind the development of commercial relations with the countries of Southeast Asia. Most spices were only harvested in the archipelago at that time.

It is not known who was the first Indian religious figure to set foot on the land of the archipelago. The discovery of Vishnou statues in Cibuaya (West Java) and in Kota-Kapur (Bangka Island) make it clear that Vaishvana teaching, which venerates Vishnu as the supreme god, was developing at this time.

The same teaching was adopted by the Mulavarman of Kutai and the Purnnavarman of Tarumanagara. Fa Hsien's narrative only refers to a few Buddhists in *Yeh-p'o-t'i*. He says that it is possible for the followers of Vaishnava/Hinduism and Buddhism teachings to coexist. Although Indian cultural elements have affected the local culture of Nusantara, it has not lost its own identity. These influences embody a new culture that is still evident today.

From the first to the fifth century of the Common Era, the commercial ports of Nusantara are close to the main sea routes of the Strait of Malacca, such as Barus on the west coast of North Sumatra, or Kedah on the Malay Peninsula.

In summary, the basis of international trade in South East Asia dates back to the first, second and third centuries. These commercial activities brought Indian cultural influence into various areas of daily life. It is clear that the people of Nusantara eventually embraced the teachings of Vaishnava, Hinduism, and Buddhism, and that the emergence of the Nusantara kingdoms was influenced by Indian culture.

Adaptation and adoption of the political system There is proof of the interactions that took place between India and Nusantara in the discovery of several archaeological sites dating back to the beginning of our era, such as the area of wetlands on the eastern coast of Sumatra, in Buni (Karawang, West Java), in Patenggeng (Subang, West Java), and in Sembiran (Bali). These findings attest to Nusantara's relationship with Arikamedu, one of the ports in southern India. The found artefacts that pay testimony to this relationship are mainly pieces of typical Arikamedu pottery and the semi-precious cornelian gemstone pearls and beads that were traded in southern India at that time (especially around the fifth and sixth centuries).

From the seventh to the eighth century, kingdoms influenced by Indian culture were established in Suvarnnadvipa (Sumatra) and in Yavadvipa (Java). The kingdoms of Srivijaya and Medang/Mataram were established by the Sailendra dynasty. As empires built on the maritime environment, the populations of these two kingdoms share a common maritime culture, although agrarian culture also developed in Medang.

The system of government in Srivijaya was composed of a datu group with its own territory. Srivijaya is therefore described more precisely as Kadatuan Srivijaya. The Kadatuan of Srivijaya was under the influence of Indian culture and the majority of the population of Srivijaya embraced the Buddhist teaching of Mahayana. The Buddhist teaching in this kadatuan was very advanced and so the Chinese monks who sought to continue their studies went to Srivijaya in order to learn the Sanskrit grammar before going to India.

The kingdom of Medang, meanwhile, consisted of a group of Rakai, with its own *lungguh* (territory). The highest ruler was called the *Maharaja*, the most famous of whom was undoubtedly Sri Maharaja Rakai Panamkaran. As for the Kadatuan Srivijaya, the people of the Medang kingdom embraced the Buddhist teaching of Mahayana. The kings who governed this kingdom built many *stupas* (relic-filled monuments), such as Borobudur, Sewu, Plaosan, and Kalasan.

The Kadatuan of Srivijaya and the kingdom of Medang possessed formidable maritime power, which is testimony to their influence across the sea. In Nalanda (India), one of the kings of the Sailendra dynasty built dormitories and *viharas* for the *samanera* – the young Buddhist monks of Srivijaya. The inscription in stone at Ligor (775) by Nakhon Srithammarat (Thailand) informs us that the king of the Sailendra dynasty built three sacred buildings of Mahayana Buddhism. Something else that Srivijaya and Medang share in common is the shipbuilding expertise known as 'sewn-plank-and-peg techniques', which were developed in all the waters of Southeast Asia. Traces of this technology can be found in Sumatra and Java, where the shape of the ships is smelted into the reliefs of the Borobudur stupa that were erected during the golden age of the Medang kingdom (eighth century). Reliefs describe at least 11 types of marine transportation, ranging from small river craft to large sailing and outrigger vessels for ocean travel.

The ships developed with these naval technologies made it possible to transport goods from various places and kingdoms. In Srivijaya's heyday, in the eighth and ninth centuries, these types of ships allowed the regime to assert its sovereignty over the seas. Regulations were issued by the Datu (elders) of Srivijaya, including the notable decree that foreign merchants wishing to trade in the sovereign territory of Srivijaya should use Srivijayan ships.

The Kingdom of Singhasari, another kingdom of Nusantara that was influenced by Indian culture, also had formidable maritime power. This regime lasted for less than a century (1222–1292), but many remarkable things were accomplished in that time. At the beginning of the Singhasari kingdom, a mixture of Hindu and Buddhist teaching, called Shiva-Buddhist teaching, was adopted by Singhasari. The development of this teaching reached its peak during the era of King Kertanagara and largely influenced Sumatra. During the Singhasari era, Indonesian ancient arts were at their peak, as can be seen in statues such as those of Prajnaparamita.

During the reign of Kertanagara (1254–1292), the idea of unifying Dvipantara (Nusantara) was carried out in order to avoid Mongolian attacks. King Kertanagara defended the idea of expanding Cakrawala Mandala beyond Java, thereby covering the entire Dvipantara region, as the Camundi inscription (1292 CE) clearly indicates. Before carrying out the idea of unification, Kertanagara led an expedition from Pamalayu to the kingdom of Malayu in the upper reaches of the Batanghari River. This expedition was a peaceful expedition, the aim of which was to restore the statue of Amoghapasa, which had been received as a sign of friendship by the king and people of Malayu-Dharmmasraya.

Written sources in Chinese, Arabic, Indian and Persian indicate that the development of navigation and maritime trade between the Persian Gulf and China since the seventh century was due to the growth of the great empires at the western and eastern extremes of the Asian continent.

At the western extremity was the Muslim Empire, dominated by the Umayyad Caliphate (660–749 CE) and later by the Abbasids Caliphate (750–1258 CE), and at the eastern extremity of Asia, a Chinese Empire that was dominated by the T'ang dynasty (618–907 CE). These two empires certainly stimulated the development of Asian shipping and commerce, but let us not forget that the empire of Srivijaya assured control of the Strait of Malacca from the seventh to the eleventh century. These empires were maritime empires focused on the development of navigation and commerce.

Religious change
The commercial links that unite the people of India with the populations of Nusantra then evolved towards a religious connection. Indian priests joined up with commercial voyages in a bid to expand their religion, and they were then followed by student priests visiting religious centres in India.

As part of this reciprocity, several communities from Nusantara welcomed the culture and religion of India. There is evidence of this close relationship in the remains of statues and inscriptions that have been found at a number of locations. The discovery of a bronze statue of Buddha *Dipangkara* in Sikendeng, western Sulawesi, indicates that a Buddhist community inhabited the area. The *Dipangkara* Buddha was considered the protective god of sailors.

Remains in the form of statues and inscriptions can also be found elsewhere, such as on the island of Borneo. Inscriptions carved on seven stone pillars were found in Muarakaman, Kutai, east of Kalimantan. The inscriptions, which were written in Pallava and in Sanskrit, probably date from the fourth century, which corresponds with the first period in which the island of Borneo was in contact with Indian culture and religions. The king who ordered these inscriptions was King Mulavarman, one of whose inscriptions describes his genealogy going back to his grandfather. Interestingly, the name of his grandfather, Kundungga, seems to be a local name that survived the influence of Indian culture unscathed. It was only during the reign of his father, Asvavarman, that the influence of Indian culture appeared and Indian names began to spread.

The religion embraced by King Mulavarman that appears on the yupa inscriptions is a Vedic religion. Older than Hinduism and Buddhism, the Vedic religion is a faith that was imported and developed by the Aryans in northwestern India between 2000 and 1500 BCE. The Aryan belief was incarnated in the sacred text of the four Vedic books, namely *Rigveda, Yajurveda, Samaveda*, and *Atharvaveda*. The followers performed ceremonies in a sacred field called *ksetra*, and not in sacred buildings or temples (*prasada*). The name of the sanctuary is Vaprakesvara, which can be interpreted as a 'mound for the god' (MacDonnell, 1954: 269). At the ceremony, alms are given in the form of 'gold in large quantities' (*bahusuvarnnakam*) and '20,000 oxen' (*vimsatirggosahasrikam*). The yupa, the post to bind the animals during the Vedic ceremonies, was an important element during the reign of King Mulavarman. The yupa was made of wood or stone and was always present in a *ksetra*. The god mentioned in the inscription of the Yupa is the god Agni (fire), because in the offering ceremony, the god Agni was considered an important witness who was to deliver the offering to the gods. The book of Rigveda also mentioned 33 other gods, incarnations of the forces of nature.

In the first century CE, a faith that paid homage to three gods (*Trimurti*), namely Brahma, Vishnu and Shiva emerged in India. This faith is in fact Hinduism. As well as these three gods, the Hindus worshipped the 33 gods mentioned in the *Rigveda*, because they still considered the Vedas their holy books. Hinduism also recognises other holy texts, the books of the *Puranas* (the ancients).

Several stone inscriptions were found on the island of Java, that is to say close to Jakarta, Bogor and Banten. These were written in Pallava and Sanskrit and date back to the fifth century. The reigning king at the time was Purnnavarman, from the Kingdom of Tarumanagara. Though he considered himself the equal of Vishnu, he never claimed to be Hindu. He was an adept of Vedicism with a cultural interest in Vishnu Vikranta or Vishnu Trivikrama. In one of the inscriptions, there is a reference to him making a gift of one thousand oxen, according to the ceremonial procedures of the Vedic religion.

The Hindu religion that pays tribute to the *Trimurti* (Brahma, Vishnu, Shiva) appeared in Java in around the seventh century. We know about the presence of Hinduism in Java from the *Tuk Mas* inscription, which was found at Dakawu, in Magelang. The inscription was carved into a stone from the river. Some of the attributes of the *Trimurti* gods were carved into it. Although the *Trimurti* was comprised of three gods, only Vishnu and Shiva had an important cult. Among Hindus, some adore Vishnu (Vaishnavism), while others worship Shiva (Shaivism). In general, it was Shiva, also known as Mahadeva or Mahesvara, who was considered the supreme god in the Trimurti. One of the architectural examples in which the Trimurti was represented, and where Shiva was glorified, is the temple of Prambanan. It consists of a complex of temples in which the great temple of Shiva is located at the centre, and flanked by two small temples, namely the temple of Vishnu and the temple of Brahma. Hinduism also developed in other islands such as Sumatra and Bali.

The presence of Buddhism in the archipelago in the second and third centuries had already been signalled by the discovery of the statue of the Dipangkara Buddha, but our knowledge of the early development of Buddhism in Nusantara is based primarily on the notes written by the Chinese voyagers who stopped there: Fa Hsien, Hwuiing, and I-Tsing. The Chinese travellers noted that the people of the 'Southern Ocean' had embraced Theravada Buddhism, and it was only in the fifth century that Mahayana Buddhism began to prosper on the island of Sumatra.

Considered to be the oldest Amaravati-style statue on the island, the discovery of the stone statue of the standing Buddha on Siguntang Hill in the region of Palembang attests to traces of Mahayana Buddhism in Sumatra. Several inscriptions have been found that date from the seventh century, at the time of Srivijaya, the oldest of which is dated 682 (CE). Following the example of the inscription of Talang Tuo, dated 684 (CE), the inscriptions clearly bear the cadences of Mahayana Buddhism. The text in the Talang Tuo inscription describes the development of the park in Sriksetra being for the welfare of all creatures, on the order of Dapunta Hyang Sri Jayanasa.

Meanwhile, on the island of Java and particularly in West Java, Mahayana Buddhism developed earlier – around the sixth and seventh centuries – as is attested by the discovery of votive tablets on the site of Segaran V, in Blandongan Temple, in Batujaya, Karawang, West Java. 43 pieces were found in the form of rectangular clay seals with rounded tops decorated by the relief of a mandala, with three Amitabha Buddhas sitting cross-legged on the upper part of the tablet, with their hands forming the *abhaya mudra* (representing fearlessness), while in the lower part, two Buddhas with their hands forming the *dhyani mudra* surround the Buddha who is sitting with both legs dangling in the *pralambhada* position. These clay seals were used in Buddhist rituals as a sign of pilgrimage when visiting holy places. Another discovery is the fragment of a terracotta inscription that contains a Buddhist spell, the ye-te, which is written in Pallava and Sanskrit.

The Buddhist religion of Mahayana propagated in Central Java and East Java since the eighth century. There are many traces of Buddhist inscriptions and monuments in the region. The earliest inscription indicating the presence of Mahayana Buddhism is the Kalasan inscription, dated 700 Saka (778 CE), which mentions the establishment of a sacred building for the cult of the Goddess Tara, the wife of Dhyani Buddha/Bodhisattva. The sacred building to the Goddess Tara has been identified as the temple of Kalasan. In addition, the Kelurak inscription, dated 704 Saka (782 CE) also mentions the establishment of a sacred building dedicated to Manjusri, the Sewu Temple. The two inscriptions of King Rakai Panangkaran were written in the Siddham/Pranagari script and in Sanskrit.

Buddhism and Hinduism will later merge in a form of syncretism. The syncretism in Java was a combination of two beliefs: of Shaivism (Hinduism) and Mahayana Buddhism (Buddhism) that merged into Shiva-Buddhism, and of Vaishavism (Hinduism) and Shaivism that came together in a school of teaching worshipping Harihara (the union of Vishnu and Shiva in one statue). The temple statues that illustrate this syncretism were generally realized during the Majapahit era (in the fourteenth century).

The idea according to which the king is the manifestation of a god reigning over the world has long been accepted, especially in Java. The king would adopt a religion reflecting his title until death, when, for his family and for the people who followed him, he then became a statue of the king, resembling the god that he worshipped, erected in a temple. This is known as *Dewaraja* worship. The cult of *Dewaraja* was born under the reign of King Airlangga (1019-1042). As a king who worshipped Vishnu, he was regarded as the incarnation. In his inscriptions, he used a royal seal, or garudamukha, a vehicle of Vishnu. After Airlangga's death, a statue was erected, later found at Petirtaan, Belahan, which represents Vishnu riding his mount, the legendary bird-like divinity Garuda. The concept of dewaraja continued and was observed by the kings of Kadiri, Singhasari and Majapahit. Thus, when he died, the first king of Majapahit, Kertarajasa Jayawardhana, better known as Raden Wijaya (1293-1309), was buried in the temple of Simping with the statue of Harihara (Vishnu and Shiva conjoined in a statue). The temple of Simping or Sumberjati is located in the southern part of Blitar, and only its foundations survive.

The fifteenth century was the period when the influence of Hinduism and Buddhism appears to fade and merge with existing local beliefs. This can be seen from the remains of the archaeological buildings scattered on Mount Penanggungan, and Mount Arjuno, etc. Here one can observe an attempt to make the mountain summit the home of the gods, or the venerated ancestors. Many sacred buildings were built with a terraced structure on the sides of the mountains, and at the top there were altars for offerings. In the mountains of Tengger lives a community worshiping a cult at Mount Brahma (Bromo), which is the dwelling-place of their ancestors, descendants of Roro Anteng and Joko Seger. Each year they organize the *Kasodo* ceremony at which they give offerings that are thrown into the crater of Bromo. This ceremony has taken place since the Majapahit era (in the fourteenth century), and perhaps even from the time of ancient Mataram (tenth century). It seems, therefore, that the animism and dynamism that had already flourished in prehistoric times were coming back to the surface.

Architecture
The only remains that have survived from ancient buildings in the Hindu and Buddhist style are from the parts that were made in stone and brick. Residential buildings were mostly made of wood, bamboo, and leaves, which easily decomposed and could not survive over time. The ancient Hindo-Buddhist buildings of Nusantara are commonly referred to as a *candi* (temple), which is derived from the word *candika*, another name for Durga, the goddess of death.

At first the temples were built for worshipping the gods, and the statues of the gods were arranged within them. However, since the emergence of the devaraja cult, the Hindu temple also became devoted to the worship of the dead, particularly of kings and eminent persons. The sraddha is a ceremony that is organized especially for kings once they have been dead for twelve years and it gives rise to the erection of a statue of the god that the king venerated during his lifetime. A temple will then have to be built in order to support its worship. The pit (*sumuran*) of the temple is then installed with a *pripih*, a stone box containing various objects such as carved pieces of gold, or silver and agate as an offering. This *pripih* is considered to be symbolic of the physical properties of the king and its contents are destined to be reunited with the god following his incarnation.

As a temporary place of residence for the venerated god, the temple is a replica of the true home of the god, which is Mount Mahameru, so the temple is therefore decorated with various sculptures made of motifs that imitate the nature of the mountain. The distinctive traits that distinguish the Hindu and Buddhist temples can be seen on the upper part of the temples. The roof of the Hindu temple is a replica of Mount Mahameru and its summit is similar to a lingga (the symbol of Shiva's man-

hood), while the roof of the Buddhist temple takes the form of a *stupa*, like the temple of Sewu, for example.

The temple, as a building, is composed of three parts: the legs, the body, and the roof. In Hindu philosophy, the three parts of the temple represent three kingdoms: the foot of the temple represents the *bhurloka*, the lowest kingdom, intended for humans, who are still bound to the world of animals and demons by their lust, and known in Buddhism as *kamadhatu*; the body of the temple represents *bhuvarloka*, the middle kingdom occupied by people who have attained enlightenment and by the minor gods, and known in Buddhism as *rupadhatu*; the roof of the temple represents *svargaloka*, the highest kingdom, occupied by the most important gods (Brahma, Vishnu, Shiva), and known in Buddhism as Arupadhatu or Nirvana.

The foot of the temple is generally square, rather elevated and may be equipped with a staircase leading to the temple chamber. In the centre of the temple there is a pit for the *pripih*. The body of the temple consists of a room containing the statue of the god/statue of the manifestation of the king, which stands in the middle of the room, just above the pit, and is turned towards the temple entrance. The exterior walls of the temple chamber consist of niches that are filled with statues. Inside the Hindu temples, the niches on the south side are filled with statues of Shiva-Guru or Agastya, the niches to the north are filled with statues of Durga, and – according to the direction in which the temple is facing – the niches to the West or East are filled with statues of Ganesh. The roof of the temple consists of a three-tier structure, which is as high as the roof is small, with its peak adorned by a kind of bell.

There are significant architectural differences between the temples of Central Java and East Java, each of which can be defined as a style. The temples prior to the year 1000 CE, including some temples of East Java, are classified in the style of Central Java. The temples built since the eleventh century (including the temples of Muara Takus and Gunung Tua in Sumatra) are classed as being in the East Java style. The most important differences between these two styles are as follows: for the Central Java style, [1] the shape of the structure is voluminous; [2] the roof has tiered sections; [3] the top is shaped like a *ratna* or *stupa*;

[4] the portals and niches are adorned with kala makara; [5] the reliefs are projected rather high in the background and in a naturalistic style; [6] the positioning of the temple, which is located at the centre of the complex; [7] they mostly face toward the East, and [8] are mainly built of andesite stones. For the East Java style, meanwhile, [1] the shape of the structure is longer and thinner; [2] the roof is a combination of several different tiered sections; [3] the top is in the form of a cube; [4] there is no makara, and only the upper part of the portals and the niches are decorated with the head of kala [5]; the reliefs are projected in a rather flat way and the images symbolically resemble the wayang kulit [6]; the layout of the temple is located behind the complex; [7] they mostly face towards the West; and [8] they are mainly built of brick.

Sculpture
Some parts of the buildings have had two-dimensional reliefs sculpted into them containing stories or simply ornaments in the form of plants, animals, and celestial creatures. There are two different *langgames* (styles) in the two-dimensional sculpture of Java. The style of Central Java gives us sculptures in relief (high relief) with a naturalistic inspiration. The high relief reveals a form of sculpture that is attached almost intact to the base of the surface of the wall. Examples of these types of reliefs are to be found in the panels of the Prambanan temple. The East Java style adopts a sculpture of bas-reliefs and symbolically resembles the wayang kulit. Examples of these types of relief can be seen in the panels of the Jago temple.

The statues often take the form of human beings or animals, or of abstract figures that resemble humans, animals, or a combination of the two, and they are made in naturalist, symbolic or schematic styles. Initially, the statuary was the incarnation of ancestral figures, or the persons worshipped by the primitive communities. When Nusantara was exposed to foreign cultural influences, especially from India, statuary began to represent the gods venerated in Hinduism and Buddhism. The statues of the gods of the two religions (and their beliefs) have distinguishable characteristics or *laksana*. The statue of the god Shiva is thus different from the god Vishnu, who is herself distinct from Dhyanibudha and the Bodhisattva. These statues generally filled the main chambers and

niches of the temple, depending on the direction of the wind. Thus, in the Hindu temple, the statue of Shiva, the principal god, was turned towards the East, while the statue of Ganesh was placed in a niche turned towards the West. In the Buddhist temple, the placement of the statue, especially the statue of Dhyanibudha, also involves the cardinal directions, as is the case, for example, with the statues of Buddha in the temple of Borobudur. In the art of sculpture, there is also talk of *langgam* or style. For example, the *langgam* of Singhasari (thirteenth century) is recognized by the sculpture of lotus trees placed on either side of the statues. On the other hand, in the language of Majapahit (fourteenth century), the lotus-tree sculpture arises out of the vases surrounding the statues.

Literature
Literary art has continued to grow since the ninth century, through interactions between the populations of Nusantara and foreign cultures, especially those of India and the Middle East. These literary works appear in the form of ancient manuscripts composed of fragile materials such as palm leaves, nipa leaves, bamboo, *dluwang* (bark), parchment, and paper. Because they were written on fragile materials, these texts were often rewritten afresh so that the content of the manuscript might survive to be understood by later generations. Consequently, not many ancient manuscripts have been found – particularly from the ninth and tenth centuries – that were written in the original script, for example, in Old Javanese. Most of the manuscripts that have been recovered have already been rewritten in a script with its origins in Bali, where people still practice Hindu customs and uphold the *Mabasan* traditions, the art of reading, and translating the old Javanese poems into Balinese.

The oldest works are the Indian *wiracarita* (heroic stories) of the *Ramayana* and the *Mahabharata*. The history of Ramayana is emblazoned on the reliefs of the panels of the temple of Shiva, the main temple in Prambanan, which was built in the middle of the ninth century. Later, the *kakawin* or long narrative poems of the *Ramayana* will be adapted to Old Javanese during the reign of King Sindok (929–947) by a kawi named Yogiswara. The work of adapting the history of Mahabharata to Old Javanese was done under the order of King Dharmawangsa Teguh Anantawikramotungga-

33

dewa, who reigned between 990 and 1016. Of the 18 *parwa* (chapters) of the Mahabharata, only nine survived, and four were completed under the reign of King Dharmawangsa Teguh, namely 'Adiparwa', 'Wirataparwa', 'Bhismaparwa' and 'Uttarakanda'. The two epics, the *Ramayana* and the *Mahabharata*, also produced other independent stories, based on events extracted from the principal texts.

Especially among the former Javanese communities, the literary arts grew and reached their peak during the Kadiri period of the thirteenth century. Many literary works were produced at that time, both in the form of *gancaran* (prose) and *tembang* (poetry). Among the most important literary works of the Kadiri period, we find, among others, the *Arjunawiwaha*, written by Mpu Kanwa, *Kresnayana*, by Mpu Triguna, *Sumanasantaka*, by Mpu Monaguna, Smaradahana, by Mpu Dharmaja, the *Bharatayuddha*, written by Mpu Sedah and Mpu Panuluh, and Mpu Tanakung's *Werttasancaya*.

Here it is worth mentioning a very famous literary work and genuine historical reference, namely the kakawin of *Nagarakertagama*, which was written during the Majapahit era of the fourteenth century. Composed by Mpu Prapanca in 1365, this kakawin is very important historically because it describes the history of the kingdoms of Singhasari and Majapahit in accordance with the inscriptions. This kakawin was also called *Desawarnana* or 'description of the various regions' because it mentioned the regions conquered by Majapahit, and the voyages of King Hayam Wuruk throughout most parts of East Java. This description was followed by an inventory of the temples, and the ritual of religious ceremonies.

It is also important to mention the work *Pararaton*. This book is a summary of *Serat Pararaton ata Katuturanira Ken Angrok*, which means 'The Book of Kings, or Tales of Ken Angrok' and it is written in the prose form of *gancaran*. This book was also conceived as a historical description, but it is very unreliable because it also contains a number of supernatural tales and passages.

Music and the performing arts
Different types of plants (including bamboo and tree trunks), earth, metals and animal parts are used in the manufacture of musical instruments.

The sounds produced by musical instruments can be classified into four categories, namely [1] idiophones, the types of musical instrument for which the source of sound is the body of the instrument itself, for example the gong; [2] membranophones, a musical instrument whose source of sound derives from materials that act like membranes, like the skin on the drum; [3] aerophones, where the sound source is provided by the blowing of air which causes reeds or columns of air to vibrate, for example, with flutes and *serunai*; and finally, [4] chordophones, which is the name for a musical instrument whose sound source is derived from plucking the strings, as with, for example, the kacapi and the *rebab*.

The earliest musical instrument in Indonesia is the nekara, the bronze drum from Dong Son that dates from the prehistoric period known as the Metal Age (about 500 years before the Common Era). These artefacts have mainly been found in the central and eastern parts of Indonesia, such as Bali and the east of Nusa Tenggara. Nekara is classified among the group of idiophonic musical instruments, because the sound is produced by striking a tympan. The centre of the striking area was usually decorated with twelve angular stars, while the edge of the striking zone was adorned with four frogs. The presence of the frogs allows us to suppose that the nekara was played during a ceremony for rainfall, the dry season lasting long enough in the east of Nusa Tenggara. Another type of nekara is the moko, which is smaller and thinner. Many mokos were found east of Nusa Tenggara. They are still used in traditional Timorese ceremonies.

The types and forms of the musical instruments are largely influenced by the penetration of Indian culture into Nusantara. There is evidence for this in the temple reliefs and in written sources, inscriptions and manuscripts. The reliefs from the temples of Borobudur and Prambanan, for example, depicted people playing musical instruments such as the drum, the flute, the sitar, and the gong. Musical instruments such as the gamelan (*tuwung*), kenong (*brekuk*), drums (*padahi*), cymbals (*regang*), gambang, saron, kacapi (*wina*), and the flute (*salukat/salgo*) are also mentioned in the tenth-century inscriptions. The musicians of Borobudur bore the status of *mangilala drawya haji*, employees who lived in the palace and were paid by the kingdom;

but there were also musicians who lived outside the palace.

Performing art is closely linked with music, which is occasionally played to accompany performances. We know that there have been performing arts in Nusantara since the prehistoric age. The performances consisted mainly of dances – war dances or dances to accompany ceremonies – both sacred and profane. The ceremony for requesting rain featured music played on the bronze nekara, accompanied by dances.

The performing arts were enriched by the propagation of Indian culture in Nusantara, but the region also retains its own traditions, such as the *wayang* spectacle, which is believed to have been in existence since the Neolithic period (Haryanto 1988: 25; Mulyono 1989: 55). The wayang, which literally means 'the shadow', was originally a shamanic ceremony designed to establish bonds with the spirits of the ancestors. When Indian culture began to permeate Nusantara, the gods of the Hindu pantheon were also considered their ancestors, such that Indian literary works such as the *Ramayana* and *Mahabharata* became the principal stories in the staging of the wayang.

Brandes listed the wayang as one of the ten elements of indigenous culture of Indonesia. The full list of these ten elements is as follows: [1] the knowledge of wayang, [2] the gamelan, [3] the metric system, [4] the art of weaving and batik [5] the metal industry, [6] the monetary system, [7] navigation [8] astronomy, [9] agriculture using irrigational systems, and [10] government agencies (Brandes 1889: 122–123).

Artistic testimonies can be seen in the reliefs from the temples of Central Java and East Java, as well as in the inscriptions and manuscripts. In the inscriptions dating back to the tenth century, the performing arts, usually performed to animate ceremonies, often present characters and high-ranking officials enjoying entertainment. Performances included *mamidu* or *mangidung* (singing), *mangigal*, *rawanahasta* (dance), *matapukan* (mask dance), and *mabanol* (comedy). In terms of the wayang spectacle, both *wayang orang* and *wayang kulit*, the pieces are created from the stories of the *Ramayana* and the *Mahabharata*. For example, in the Wukajana inscription, dating from 908, it is written that 'the Galigi manipulated or played the wayang for Hyang (God) with the story of

Bhima Kumara'. The story of Bhima Kumara, or Bhima as he was known when he was younger, was taken from the *Wirataparwa*, which is part of the *Mahabharata* epic. On the other hand, it is also mentioned that 'the Jaluk relates the *Ramayana* (history)', which refers to the *mabasan*, a tradition still practiced in Bali, at which people gather to listen to the reading of the kakawin and its interpretation in the Balinese language.

INTERACTIONS BETWEEN NUSANTARA AND CHINA

There have been maritime trade relations between Nusantara and China for a long time. China already had a trade network with other regions in the west and the north, and the land route for the silk trade has been known about since the period of the Han dynasty (206 BCE–220 CE). This route began in Xi'an in China and then crossed Gansu, before entering Central Asia in Turkestan, and heading on towards the Mediterranean and Europe.

The other route is the maritime Silk Route (or Silk Road), which is believed to have developed at much the same time as the land route. Later, many spices from Indonesia and ceramics from China were transported via this sea route, which is why it is often referred to as the Spice Route or the Ceramic Route. It was an international trade route from southern China to Indonesia that passed through the Straits of Malacca and headed on to India, before the road then split in two: the first branch crossed the Persian Gulf and went on by land into Syria and the Mediterranean; the second crossed the Red Sea before reaching the Mediterranean. The relay was carried out by Venetian merchants who took care of distribution in Europe. According to historical documents, the Silk Road was well known to local communities. They used the route for commercial purposes, particularly for transporting spices from the Moluccas. Between 138 and 126 BCE, Emperor Han Wudi of the Han Dynasty (206 BCE–220 CE) ordered his servant Zhang Qian to lead an expedition in search of a route to the West (Europe). This route became a leading trade route, as well as a route for cultural propagation. The route is also known for having served as the gateway for the spread of Buddhism (Li Zhiyan, 1989, Suropati, 2016).

The Indians were aware of the location of Nusantara and Svarnnadvipa (another name for

Sumatra that means 'the Golden Isles') from the beginning of the Common Era until the third century. Relations with China came later, as evidenced by the Chinese chronicle originating from the travel report of the Buddhist monk Fa Hsien (Fa Xian) and Gunavarman (see above).

With Palembang at its centre, the kingdom of Srivijaya was known to the Chinese as early as the seventh century, as is clear from the notes of I-Tsing, a Chinese Buddhist monk who was on his way to India to spread the word of Buddhism. His notes teach us more about the role of ships from Srivijaya and Malaysian crews travelling to China. I-Tsing left China for Nusantara by embarking on a Persian vessel, before taking a ship from Srivijaya to India.

At the time of his visit to Srivijaya, I-Tsing was able to get to know the local products, such as the turtle skin, ivory, gold, silver, incense, camphor, dammar resin, and pepper that were bought by foreign merchants, or traded for ceramics, cotton and silk. According to I-Tsing, there were two kingdoms in Sumatra. In the north was the kingdom of Melayu, whose capital is the town of Jambi, located on the site of the Batanghari River, while the south was occupied by the kingdom of Srivijaya (Che-li-fo-tse and later San-fo-tsi), with Palembang as its capital. The Chinese chronicles described how the kingdom of Srivijaya succeeded in maintaining control of Malayu, the ports and regions of Sumatra, and the Strait of Malacca. Since it was the obligatory route for all foreign ships leaving India for China, and vice versa, the control of international trade and navigation in the Strait of Malacca proved highly profitable in terms of the taxes received from passing ships. Local merchants brought their products to collection points (warehouses) in the Strait of Malacca from where they were shipped to other countries. The capital of the kingdom of Srivijaya was the starting point for gathering together products and the hub of international trade.

Trade was usually organized by means of bartering, but also through the use of exchange rates. The goods that were initially involved in this global trade were silk, spices and ceramics. The strong demand for spices and ceramics at the time was to have consequences for international relations. The spices were sought by various nations for use in seasoning and medicines. The Chinese appreci-ated the cloves, which were used for their medicinal virtues and for clearing the bronchial tubes. During the period of the Tang Dynasty (618–960), cloves were used in the blending of flavours. In Europe, cloves were also used for medication and as an orexigenic foodstuff, intended to stimulate the appetite.

Nusantara was at that time the largest producer of medicinal spices and seasonings. Cloves (Eugenia aromatica), nutmeg (Myristica fragrans), pepper (Piper nigrum), cardamon (Elettaria carda-momum), candlenut (Aleurites moluccana), keluak (Areca catechu), cinnamon (Cinnamomum zeylani-cum), ginger (Zingiber officinale), and turmeric (Curcuma domestica) were exported to China, India, Western Asia, the Middle East, and Europe.

Cloves came from the north of the Moluccas, which was known to be the only place that produced them. Cloves were introduced into Europe as early as the sixth century and this spice was the most highly valued of all products, even reaching as high as the price of gold. Nutmeg was produced in the island of Banda in the Moluccas. The merchants in the Moluccas did not expect buyers to come to them, but they were also active in supplying cloves and nutmeg to Malacca. Similarly, traders from Java or Sumatra came to pick up the products from the Moluccas and transport them to Sumatra. Pepper actually came from Kerala in India, but was grown in Java around the ninth and tenth century due to the growing demand for this product on the world market. Pepper was planted in Sumatra, particularly in Lampung, Palembang, Bengkulu, and Jambi in around the fifteenth and sixteenth centuries. Indian and Chinese merchants came to pick up the pepper that was produced in Java and Sumatra and from there they took it on to India and China.

There has been a shift in political power since the end of the fourteenth century, with the Buddhist monarchy of Srivijaya giving way to Islam. Towards the end of the thirteenth century, the Sultanate of Samudra Pasai was replaced by the Sultanate of Aceh Darussalam. Malacca, which was dependent on Srivijaya, was then at its peak. The existence of state-of-the-art maritime trade led to the development of Islamic maritime kingdoms and the emergence of their capital cities and international ports such as Pedir, Malacca, Demak, Cirebon, Jakarta, Banten, Gresik, Japara, Pekalongan, Lasem, Demak, Kudus, Tuban, Surabaya, Hitu, Ternate, Tidore, Banda, Gowa-Makassar, Banjarmasin, Lampung, Jambi, Palembang, Minangkabau, and others. More traders arrived from Sri Lanka, Gujarat, Bengal, Cambay, Coromandel, Turkey, and Egypt, followed by the arrival of Europeans from Portugal, Spain and the Netherlands. The Chinese exported their quality products, such as silk and ceramics and other goods such as bronzes, cottons, woven fabrics, lacquer, lychees, tea, sugar, and minerals. Rice was also exported from China and from Champa. Vietnam exported gold boxes, pearls, scented wood, textiles, ceramics, and other products. Clothes with floral prints, white cotton, rattan and coconut leaves, gold and bronze bars, aromatic plants, medicinals, various scented woods, areca nuts, ebony, beeswax, and kapok were exported from Champa. The ceramics of Thailand and Japan were also highly prized. From Persia came boxes made of silk embroidery, rosewater, ceramics, glass and gold, ceramic wall tiles, cobalt blue minerals, and ceramic glazes. Glass boxes and anise (cough medicine) were imported from the Middle East. Pepper, silk, cotton, sandalwood, precious stones, and perhaps gold jewellery came from India. Sauces with myrrh and incense, roots scented with putchuk, and a kind of pepper (cubebs) were imported from Arabia. Cinnamon came from Sri Lanka, coffee from Yemen. Lions, anise (for cough medicines), and putchuk roots were sent from Africa.

Coastal communities are inclined to remain open and proactive in their interactions with the marine environment and it may have been this factor that allowed them to accept new cultural elements from outside. At the same time, these elements, such as ornaments, were adopted and developed through their adaptation (acculturation) to local traditions, or to meet local needs. The use of Chinese ornamental elements, on both ceramics and silk, can be seen on batik fabrics, wood carving, lacquer and glass painting in Indonesia. According to Zhu Fan Zhi's business reports, which were written by Chau Ju-Kua in 1225, during the heyday of the kingdom of Sriwijaya (from the seventh to the fourteenth century), the Chinese used tropical products such as benzoin, scented woods, camphor, spices, handicrafts, and valuable objects such as pearls, ivory, rhinoceros horns, and animals that

were only native to Indonesia. Products arriving from China generally included silk, brocade, metal products, coins, crafts, and ceramics.

The arrival of Chinese cultural elements to Indonesia came about through the international activities of maritime trade. The most important Chinese products were silk and ceramics. It is believed that the tradition of silk manufacture began during the Shang dynasty (around 1400 to 1300 BCE). Ceramics, particularly terracotta, began to be made in China during the Neolithic period, or in about 8,000 years BCE, while lacquered objects began to be produced around the period of the Shang dynasty (1600–1046 BCE). In Indonesia, it seems that lacquers were produced only in Palembang, in the south of Sumatra. They are called *lak* in Palembang, which is derived from lacquer in English, an extension of the word lac, which is used to describe the resinous material produced by an insect called *Laccifer lacca*. The plant is widely distributed in Japan, China, and the Himalayas. In Indonesia, especially in southern Sumatra, the plant is referred to as the kemalo tree.

The manufacture of art objects through the use of silk, ceramics, lacquer, and wood has a long and important history in China. Art in China, which is handed down from generation to generation, is also a means for expressing one's beliefs through various forms of ornaments, such as dragons and birds. In China, the dragon is a symbol with many facets. It is a very complex, sacred and mythical animal generally considered to be the symbol of fertility. Since the period of the Han dynasty (206 BCE–220 CE), the five-clawed dragon became the symbol of the emperor, who was regarded as the son of heaven. When depicted with a dragon, the Hong (huang) bird symbolises the union of an empress and an emperor. Feng, which means male bird, embodies the cosmological concepts of the universe or yang, while huang represents the attributes and characteristics associated with the female. The ying, represented with peony flowers, refers to the love for one another. Lasem Batik, for example, which is decorated with the hong-bird motif (National Museum of Indonesia, inv. 29040), is comparable with the blue-white ceramic dish. It seems that the manufacturers of Lasem Batik were attempting to represent hong birds in different styles.

There are Chinese elements to be seen in almost all parts of Indonesia, be it in decorative patterns or colour schemes that have not simply been copied, but also adapted. These techniques were generally found in coastal areas where communities were more open, as in the case, for example, of the batik manufacturers of Cirebon. One might imagine that Chinese motifs in the form of clouds and corals were easily accepted, because by living in coastal areas near the sea, the Indonesian communities were used to observing natural phenomena such as *mega mendung* (clouds) in the sky, or *wadasan* (coral) in the sea. A visit to Cirebon provides the opportunity to enjoy the wonderful natural landscapes, such as the panorama from Mount Ciremai, which is located inland. It is assumed that there was a perception of motifs that were shared in common between the Chinese and the people of Cirebon. The *mega mendung* (clouds), for example, are patterns, usually painted, that symbolize the higher world associated with rain and with fertility. As well as the Chinese elements, Islamic and Hindu elements were also absorbed at Cirebon. The Hindu elements were in the form of mountains (*gunungan*) and the community of Cirebon used Islamic elements in the form of Arabic letters in a number of distinctive forms that were different from the other decorative arts.

Chinese culture was therefore to have a great influence on the traditional fabric of Sumatra, especially in Bangka, Palembang, Aceh, Lampung, and Minangkabau. Chinese cultural elements played an important role in the development of the visual arts in Indonesia, albeit to a lesser extent than the Hindu-Buddhist and Islamic cultures. Many forms of Chinese artistic expression were adapted into local artistic forms. Particularly in the coastal regions of Cirebon, Pekalongan, Lasem, Demak, and Kudus, therefore, the batiks of Java are adorned with motifs such as flora, fauna and other natural phenomena from China. The decorative motifs of the flora included lotus and chrysanthemum; the ornamental motifs of fauna consisted of fantasy animals, such as dragons and hong birds; and the motifs depicting natural phenomena included clouds, coral, and mountains. The colouring used in Chinese culture was generally red and yellow, including the use of gold thread for textiles (Subarna, in Chambert-Loir, 1990).

The process of cultural exchange

It is hard to know the precise date of Islam's entry into Nusantara, especially since trade relations between the kingdoms of Nusantara and the Middle East, Persia, India and China are so long established. It is highly likely that it was merchants from Persia who first brought Islam with them to Nusantara.

There are several theories on the subject. Islam is claimed to have come from Arabia, Persia, India, and some even say that it spread through China. Although opinions on the origin of Islam may vary, they all agree on the role of the merchants. This is all in keeping with the hadith that reads: 'Convey from me, even if it is a single verse.' When Islam came to Nusantara, it therefore spread via the local ulama, or through apostles of Islam like the nine apostles (or Wali Songo) from Java.

In the search for the origins of Islam's penetration, we are able to know about the form in which it was introduced to Nusantara. For example, during the era of the caliphate that followed the death of the Prophet Muhammad, one of the great branches of Islam that started to develop was Shiism. Throughout the region, but particularly in Iran, Shiite traditions began to take root. In Nusantara, the commemoration of the day of Ashura was presented in the form of banquets and the processions of the tabot (which symbolizes the coffin of Imam Hussein) that took place in several places on the west coast of Sumatra and were celebrated by cries of '*Hayya Hussein! Hayya Hussein!*'

In Aceh, the arrival of so many merchants made relations between the Kingdom of Samudera Pasai and Persia particularly significant in the thirteenth century. At one point, one of the advisors to the kingdom of Samudera Pasai was actually Persian. The tombs of Aceh's leaders were engraved with Shiite verses written in Kufic, the script developed in the port of Kufa, in Iraq.

Indirectly, China also introduced Islam to the local population, with its main aim being that of seeking a new commercial outlet. As well as contributing to the expansion of the Kingdom of Srivijaya in the seventh and eighth centuries, the competition between the T'ang dynasty in China and the Umayyad Caliphate also increased the use of sea routes and commercial routes through the Malacca Strait and the west coast of Sumatra. The Chinese chronicles from the period of the T'ang dynasty represent the oldest sources attesting to the presence of Arab traders in Nusantara, and there is mention of the people of Ta-Shih cancelling an attack on the Kingdom of Ho-ling (Kalingga) in 674.

Groeneveldt portrayed the Ta-Shih as Arabs who had settled in colonies on the west coast of Sumatra. According to Groeneveldt, Muslim communities already existed in Canton (Tjandrasasmita, 2000: 17). Azyumardi Azra also describes how we know about trade relations between Arab peoples, Persians, and the Kingdom of Srivijaya in the eighth century thanks to two letters from the King of Srivijaya that were sent to the Umayyad caliph, Muawiyah bin Abi Sufyan (661–680) and Umar ibn Abd Al-Aziz (717–720), which contained gifts offered as a sign of friendship (Tjandrasasmita, 2009: 73). The earliest evidence for the existence of Muslim communities was discovered in Java. It is the tombstone of Fatimah binti Maimun bin Hibatullah (1082), in Leran in Gresik, East Java. Meanwhile, the headstone in Barus of Tuhar Amisuri (1205–1206) came to be seen as formal proof for the existence of Muslim communities on the west coast of Sumatra.

While it was commerce that first introduced Islam into society, other channels carried it forward, such as marriage, education, Sufism (tasawwuf) and art. The process of Islamization was facilitated by its non-coercive approach, the absence of castes within the social strata, and initiation ceremonies for embracing the religion.

The introduction of Islam to Nusantara also meant the spread of Muslim culture. The teachings of Islam were adapted to make them more easily understood by the local society, as was the case with Sunan Kalijaga who used shadow puppets as a means of teaching. Stories of Hindu-Buddhist origin were adapted to fit Islamic teachings. It is not surprising, therefore, that the Islamic culture in Nusantara is quite different from Islamic culture in its country of origin.

Moreover, the ban on representing living beings led to a modification of Hindu-Buddhist sculptures. Originally, the motifs used in engraving only allowed floral, geometric forms and depictions of landscapes. The seventeenth century finally saw the development of representations of living beings, such as humans and animals, but by disguising them in the form of intertwined leaves, or calligraphy, such as can be seen on the relief panel at the mosque in Mantingan in Jepara, Central Java.

Several centres within the Islamic kingdoms of Nusantara took the initiative of issuing their own official currency as a sign of their legitimacy. The dirham – a gold coin issued by the sultanate of Samudra Pasai – is the oldest currency from the Islamic period to have been issued in Nusantara. The traders of Pasai also introduced the forging of gold coins in Malacca in 1414, when Parameswara converted to Islam at the time of his marriage to a Princess from Pasai. The Pasai dirham was subsequently introduced into the sultanate of Aceh following the conquest of Pasai in 1524, at the time of the reign of Sultan Alaudin Riayat Syah. This was a conquest that was facilitated by good relations between the Sultanate of Aceh and Turkey, which sent experts in weaponry and forging coins. The sultanate of Palembang had two types of coin, the piti buntu (coins with no hole), and the piti teboh (coins with a hole in the centre). The shape of the piti teboh was probably influenced by the shape of the picis, the Chinese money made of copper that was in wide circulation in commercial centres at the time. The Sultanate of Gowa, in Sulawesi, also issued its official gold coin called the jingara. The kingdom of Wolio (Buton) issued a form of currency made of cotton called the kampua (cat. 118, 167–170), as did the kingdom of Maluka in Kalimantan.

Prior to the eighteenth century, the architecture of the mosques of Nusantara is characterized by a staged roof supported by four posts, such as the Menara Kudus mosque, for example, which also has a Hindu-style minaret. The use of domed, semi-circular roofs, or ogees over the door, over the window, or between the pillars of the veranda and the entrances only began to establish themselves in around the eighteenth century. These architectural elements are also present in the palaces of the Islamic kingdoms.

The Islamization of Nusantara not only included religious instruction, but also the introduction of writing and language, which can be seen on tombstones, coins, seals, inscriptions, and ancient manuscripts. In addition to the Arabic language,

Arabic writing also evolved and adapted to the languages of the local communities, such as, among others, the Malay community of Sumatra. Arabic inscriptions without punctuation (commonly known as Arab Gundul) are written in Jawi. In Java, inscriptions mixing Arabic with the Javanese or Sundanese languages are written in Pegon. The earliest inscription yet found from the Islamic period in Indonesia is on the headstone of Fatimah binti Maimun bin Hibatullah in Leran, Gresik, in East Java (1082).

Inscriptions dating from the Islamic period are generally written in a beautiful stylized form of calligraphy called khatt. The calligraphy contains quotations from verses of the Quran, hadiths, texts giving praise to Allah, shahadah, zikr, and poetry, as can be seen on the decorative motifs used on textiles, such as the fabric Basurek from Palembang. The seal of Aceh referred to the names of nine sultans and the year of 1296 in the Hijri calendar (1879 CE).

The role of Muslim, Arab, Persian, Indian, and Chinese merchants led to the growth of Muslim cities in Nusantara and Southeast Asia in the period leading up to the emergence of the Islamic kingdoms. Thanks to the tombstone of Malik al-Shaleh (1297), which mentions the saga of the kings of Pasai, we know that Samudra Pasai was the first Islamic kingdom of Nusantara. The emergence of Samudra Pasai as a kingdom could not be separated from the political situation of the Kingdom of Srivijaya, which began to decline in the twelfth and thirteenth centuries. In addition, the Pamalayu expedition (1275) led by the kingdom of Singhasari accelerated the decline of the kingdom of Srivijaya and allowed the subordinate areas to break away from its power. The Islamic kingdoms of Nusantara, such as Aceh, Malacca, Demak, Cirebon, Banten, Ternate-Tidore, Gowa-Tallo, Banjar, Kutai, and Mataram then continued to grow and develop until the sixteenth century. These kingdoms remained active in maritime and commercial transportation until the eighteenth century. They continued to structure themselves around a city port that assumed the role of the capital of the kingdom. The glory of the Islamic kingdoms in Indonesia The period classified as Islamic began in the thirteenth century and coincided with the creation of the Sultanate of Samudra Pasai in the north of Sumatra. Shortly

after his accession to the throne in 1285, the first king, King Meurah Silu (died 1297) converted to Islam, adopting the name of Malik Al-Saleh. This kingdom would shortly become a major centre for trade and for the propagation of Islam in Indonesia, particularly in the western part of the region, and its status as a centre of Islamic power is one that is to last a long time.

In 1400, another Islamic kingdom, Malacca, was established in the Malay Peninsula. Until it fell during the Portuguese attack of 1511, Malacca continuously worked together with Samudra Pasai to propagate Islamic culture. The use of Malay in both of these kingdoms as the lingua franca for trade and as the language of instruction in Islamic educational institutions led to Malay becoming the language of science, religion, and literature.

When Samudra Pasai began to decline, a new Islamic kingdom, Aceh Darussalam (1516-1700), was established not far away. Aceh Darussalam is considered to be the largest Islamic kingdom in Southeast Asia. The capital of Kutaraja not only became the kingdom's capital, but also the base of the first Islamic university in Southeast Asia. As a centre for study and educational activities, the Islamic literature of Malay flourished in Malacca until its conquest by the Portuguese in 1511. The Portuguese sovereign then made Malacca a centre for the spread of Catholicism.

In the meantime, other Islamic kingdoms appeared in the province of Riau in the period around 1512-1515, including Siak, Kampar, and Indragiri. The influence of Islam increased within these areas, as the three kingdoms had always conducted commercial activities with Malacca. The presence of Islam in Jambi has been attested since the ninth century, but Islamization on a massive scale coincided with the growth and development of the Islamic kingdom of Jambi in around 1500, during the reign of Orang Kayo Hitam. The last sultan of Jambi was Sultan Taha Saifuddin, who was killed in a great battle against the Dutch colonialists on 1 April 1904.

The spread of Islam on the northern coast of the island of Java led to the growth of Islamic kingdoms such as Demak, Cirebon, and Banten. The second King of Demak, Prince Sabrang Lor, led an attack with his fleet on Malacca. The kings of Demak were well known as religious patrons and so they shared a close relationship with the ulama (scholars), especially with the revered mystics known as the Wali Sanga. The king of the Islamic kingdom of Cirebon, Syarif Hidayatullah, was also one of the Wali Sanga and was celebrated under the title of Suhunan Jati or Sunan Gunung Jati. He also earned the nickname of Pandita Ratu ('the wise king') by spreading Islam in the country of Sunda and assuming the office of the head of government. The sultanate of Cirebon, which was a major religious centre until 1681, slipped into decline as the result of colonialism.

At that time, prior to 1525-1526, the region of Banten was still under the reign of Sunda Pajajaran, which was centred on Bogor. Banten had been an important port since the fifteenth century and was part of the trading network of the Silk Road. In 1526, Banten was conquered by Islamic forces from Cirebon. This led to the foundation of the Sultanate of Banten, which reached its peak under the reign of Sultan Ageng Tirtayasa. The Sultanate prospered in terms of economy, trade, politics, religion, and culture. However, the intervention and conquest of the Dutch East India Company stifled the Sultanate of Banten, which was abolished by Governor General Daendels at the beginning of the nineteenth century.

Particularly on the island of Lombok, the Islamic kingdoms of Nusa Tenggara spread their authority in several directions, towards such destinations as Pejanggik, Sekotong, and Bayan. In 1673, all of the Islamic kingdoms of Lombok were transferred to Sumbawa with the support of the Sultanate of Gowa, in the south of Sulawesi. The Sultanate of Bima in Sumbawa enjoyed a very close relationship with Gowa, but this kingdom was eventually destroyed under pressure from the Dutch East India Company, which constantly interfered with the royal government to impose its authority, even to the point of arresting kings that showed resistance.

The most important Islamic kingdoms of the Moluccas were Ternate and Tidore. The King of Ternate, Bern Acola, had converted to Islam and used the title of sultan, while the other kings still used the title of king, or kolano. In 1535, Sultan Hairun from the Sultanate of Ternate succeeded in uniting the zones of the northern Moluccas. Regional unity began to peter out, however, with the arrival of the Portuguese and the Spanish and their attempts to monopolize the spice trade. The

Portuguese merchants concentrated on Ternate, while the Spanish merchants looked towards Tidore. Sultan Baabullah continued to reign, proclaiming himself the sovereign of the Moluccas region and expanding his territory to Mindanao. At the beginning of the seventeenth century, the Dutch East India Company succeeded in extending the range of its political influence and monopolizing the spice trade.

However, the kingdoms that developed in the south of Sulawesi, including Gowa, Luwu, Bone and Wajo, opposed this domination. The Kingdom of Gowa was transformed into an Islamic kingdom on 22 September 1605. At the same time, it extended its political expansion into other kingdoms. Led by Sultan Hasanuddin, The kingdom of Gowa was not intimidated by the army of the Dutch East India Company, assisted by Arung Palakka, from Bone, who wanted to monopolize the spice trade. This war came to an end with the Treaty of Bongaya in 1667. The Kingdom of Gowa, in alliance with the kingdom of Wajo, suffered defeat in 1670 and King Arung Matoa lost his life during the battle. His successor, Arung-Matoa, was forced to sign an agreement in Makassa for the kingdom of Wajo to be transferred to the Dutch East India Company.

In Kalimantan there were also a number of Islamic kingdoms, both large and small. The three major kingdoms were the kingdom of Banjar in the south of Kalimantan, the kingdom of Kutai Kartanegara in East Kalimantan, and the kingdom of Pontianak in West Kalimantan. During the reign of Sultan Mustain Billah in the early seventeenth century, the kingdom of Banjar or Banjarmasin was able to raise about 50,000 warriors and was therefore feared by the neighbouring kingdoms. The power of the kingdom of Banjar was able to withstand the political influence of Tuban, Arosbaya, and Mataram. In 1607, the Dutch merchant Gillis Michielse-zoon arrived in Banjarmasin and a new enemy appeared in the form of Dutch attempts to bring about a commercial monopoly, which resulted in the confrontations led by Prince Antasari between 1859 and 1863.

It is estimated that the kingdom of Kutai Kartanegara embraced Islam in around 1575. The new Sultanate then began to spread Islam in the surrounding areas through until the early seventeenth century, when the merchants of the Dutch

East India Company began to impose their influence, although the sultanate continued to exist.

Aji Muhammad Salehuddin II was crowned the Sultan of Kutai in 2003 and still holds power today. Islamic kingdoms have developed in the meantime in West Kalimantan, including those of Tanjungpura and Lawe. These kingdoms have established relations with Java and Malacca. In fact, Islam has been present in the coastal region of West Kalimantan since the fourteenth century. Similar gravestones have been found at Tralaya and Trowulan that date back to the fourteenth and fifteenth century.

THE INFLUENCE OF CHINA, INDIA AND ISLAM ON THE TRADITIONAL FABRICS OF INDONESIA

Indonesia is one of the richest countries in the world with more than 300 different ethnic groups, multiple languages and countless dialects. The island of Sumatra is one of the largest in Indonesia and it was home to several tribes. From the seventh to the twelfth century, the local people and their culture were influenced and enriched by many foreign cultures including those of China, the Muslim world, and India, and the arrival of European countries such as England and the Netherlands in Sumatra also had an impact. These influences can be seen in the region's traditional fabrics. Each region or ethnic group in Sumatra has their own particular weaving techniques and patterns. For example, Aceh and West Sumatra are known for their fabrics that use embroidery techniques influenced by China and Europe. The calligraphic fabrics of Palembang and West Sumatra using silver or gold threads were influenced by India and China, while Lampung, which faces the tip of the island of Java, was also strongly influenced by China, India, the Muslim world and Europe.

Embroidery using gold or silver threads is a well-known tradition in the regions of Malaysia, such as Perak, Johor, Pahang, Malaysia, as well as in regions of the kingdoms or the sultanates of Sumatra, particularly in Aceh, Jambi, Lampung, and Palembang. The embroidery of Sumatra is famously beautiful and the women of Sumatra are undoubted experts in gold and silver embroidery.

Located at the northern end of Sumatra, Aceh is an area so strongly influenced by Islamic culture that it is known as Serambi Mecca (the Veranda of Mecca). This can be seen on the ornaments, patterns and designs that reveal the importation of Islamic culture in the form of plants and natural features, such as the moon, the clouds, and the waves. According to Islamic teachings, it is forbidden to incorporate elements or patterns of human or animal figures, although animal forms may still be found in certain designs, albeit in a modified or stylized way.

Sumatra is not only famous for its embroidery, but also for its songket. Songket is a fabric woven with gold or silver thread. There are two types of songket, namely songket balapak, in which the gold or silver threads take up the whole of fabric, and songket babintang (batabur), where the use of gold or silver thread only extends to certain parts of the fabric. Generally, the fabric woven with gold or silver thread is seen in the form of sarongs and scarves. The region of Western Sumatra that is famous for the production of songket balapak is Pandai Sikek, although its origins are in Koto Gadang. The songket balapak of this region is famous for its complex design and delicate manufacture, which makes it very difficult to imitate.

The kind of pattern known as Minangkabau songket bears the meaning, message and philosophy of the life of the people of Minangkabau, contained in the idea of 'alam takambang jadikan guru' (nature is the best teacher). These motifs are usually found in various rumah gadang (large house) carvings, such as the geometric patterns of the natural environment, like fern tendrils (kaluak paku), the tip of bamboo shoots, and various fabric fringes with decorative patterns known as itiak pulang patang (ducks arrive home). All these decorations have symbols or meanings that reflect the way of life of local people.

Trade relations with Asian and European peoples have had an impact on the cultural elements of Lampung. The Hindu influence can be seen in the image of the Garuda, which is, according to mythology, the chosen mount of Vishnu, the god who preserves the universe. Garuda also symbolizes the Upper World, unlike its sworn enemy the dragon, which is, in mythology, a symbol of the Lower World, and an element derived from Chinese culture. Sometimes we find the influence of local motifs, such as the motif influenced by Javanese culture that is found on fabrics depicting ships, which shows the motif of ancestors in the style of a marionette, wearing a kris (dagger) at the waist. The arrival of the Dutch East India Company in search of pepper in the seventeenth century also inspired the community to create the pattern of a ship on its traditional fabric.

The Indian influence in Nusantara can be seen in the fabric patola, which is the superior form of double woven ikat fabric from Gujarat, Western India. Patola fabric consists of a chain of highly resistant colours and the cloth is used as a royal garment, a ceremonial garment, and a costume for ritual dance (in a dance to summon rain in Tanah Ai, a village in the region of Sikka Regency, Flores, in the east of Nusa Tenggara). Historically, patola fabric began to be produced in around 700. We know about the particular technique from a chronicle dating back to 1200, but the name of the material was not known. Another chronicle by Duarte Barbarosa that is dated 1500 featured a response saying that the patola fabric sold in Southeast Asia and Nusantara was much appreciated. Another chronicle also mentioned that patola fabric was one of the principal products for export and trade in Southeast Asia in the sixteenth and seventeenth century.

Portuguese priests distributed patola cloths on the Solor Islands, Banda, the Moluccas, and Makassar through Dutch traders. Because of its sacred, spiritual value, patola fabric was transformed into clothes by the Brahmans and Jains, the priests who led the ritual ceremonies. Patola fabric was also significant on the grounds that the patterns woven into the fabric were considered to guarantee happiness and good fortune, and to prevent against disaster. The patola fabric was preserved as part of a magical or sacred heritage, and was used in ritual ceremonies related to marriage, but also to death, in which it was used to wrap the corpse.

In India, a patola garment was worn in a ritual ceremony by women who were seven-months pregnant, in order to bring the baby an aura of happiness at the time of birth. Nobles wore patola as a robe, for its spiritual value (religious magic) and for its promise of protection and well-being.

In Bali, the use of patola sheets is included as part of a ceremonial custom, in which the cloth and thread of the patola are also considered sacred. It is then reduced to ashes and mixed with drugs, because it is considered to have magical powers capable of curing madness or paralysis. In Indonesia, as well as eastern Nusa Tenggara, patola fabric is also common in Pontianak, Gorontalo and Bentenan de Manado. Patola is also known as jelamprang or cinde in Java, and as sembagi cloth in Palembang.

Coastal zones were the entry points for different cultures. In the case of batik, this meant that different patterns had an influence, hence the name pesisir batik (coastal batik). Pesisir batik was influenced by Chinese culture. The patterns are freer, symbolizing the egalitarian life of coastal communities. They are mainly characterized by the use of images of flora and fauna, such as land and sea animals, trees and leaves. Coastal areas producing batik include Cirebon, Pekalongan, Lasem, Tuban, Bengkulu, and Madura. The batik produced is characterized by bright colours, such as red, green and blue, and the use of varied patterns.

Hindu influences appear in motifs such as those featuring Vishnu and Garuda. The influence of Islam can be seen in the geometric patterns and the avoidance of realistic depiction of animals, while the Chinese influence is evident in the motifs featuring the lotus, storks, and centipede. The motif of a ship would have been the result of the influence from the Dutch. Batik from Tuban in East Java has also been influenced by Chinese culture, which appears in the blue silk pattern and hong birds of Lok Can batik. Hindu and Indian influences appeared in the kawung designs, namely panji ori, panji puro, and panji serong, while the influence of Islam appeared in the geometric form of kijing miring. At that time, batik from Bengkulu on the west coast of Sumatra was most affected by the influence of Islam. The Arabic alphabet of Jawi/Malay appeared in the fabric and was beautifully painted in the style of calligraphy to evoke besurek, which means composition or writing.

< 21. Dipankara

2nd–5th century (?)
Karama (river), Sikendeng (village),
Sampaga (district), Mamuju (kabupaten),
West Sulawesi
Bronze, 75 x 42 x 18 cm
MNI 6057

Whereas Maitreya is the Buddha of the Future, Dipankara is one of the Buddhas of the past. He is the protector of sailors and fishermen. Although his wrists and hands are missing, it is thought that the figure's right hand was in the *abhaya mudra* position ('absence of fear'), and that the left hand held the thin, regularly pleated garment that lies over the figure's left shoulder. The style of this garment is inspired by an art that developed from the 3rd century in Amaravati monastery in South India. However, the form of the ringlets of hair and absence of an *urna* (a dot or spiral between the eyebrows) is related to the style that developed in Anuradhapura, the ancient capital of Sri Lanka. This statue may have been brought by a Buddhist sailor who stopped off on the west coast of Sulawesi during a journey to the Maluku Islands in search of spices.

> 22. Standing Buddha

Lombok, Timor, West Nusa Tenggara
Bronze, 22.5 x 10.5 x 5.5 cm
MNI 8524

This statue in the *tribhanga* position (the equivalent of the Western European *contrapposto*) is in good condition, lacking only the feet and ankles. While the Buddha's left hand holds his robe, the gesture of his right hand, his thick neck and square face are all characteristic features of Javanese bronze statues. This sculpture was one of the first Hindu-Buddhist discoveries to be made on the island of Lombok, in 1960. Although the island was influenced culturally by Bali, which is close to Java, many vestiges of Indian culture have been found on the neighbouring island of Sumbawa.

< 23. Standing Shiva Mahadeva

9th century
Palembang, South Sumatra
Bronze, 85 x 36 x 22 cm
MNI 6031

Shiva Mahaveda, the supreme god of Hinduism ('Mahadeva' means 'great god'), is known as the 'destroyer of evil and the transformer'. Here he is seen with a third eye on his forehead and a headdress in the form of a crown adorned with a skull and a crescent. He wears a tiger's skin and a sacred thread in the form of a serpent. He usually carries a trident, a string of beads, a fly-swat and a jug containing the water of life. Shiva's companion is the goddess Parvati and his mount is the *nandi* (bull), though neither is included in this statue. Here, Shiva's right hand is in the *vitarka mudra* position, symbolising teaching and discussion. The bulge at the base of the statue was probably used to fix it to a plinth. The figure's headdress and clothing are influenced by the style of the sculptures linked with the Shailendra dynasty (see cat. 33).

<< 24. Prajñaparamita

13th or 14th century
Gumpung Temple, Muara Jambi,
Jambi province, Sumatra
Stone, 83 x 70 x 57 cm
Muara Jambi Temple Compound Museum,
inv. GP/II/LL/1/78

In 1978, during restoration works, this statue was found to the left of the entrance of Gumpung Temple (Hardiati 2002: 140; Reichle 2007: 65). The temple stands in a complex of over 82 ruins of religious buildings in Muara Jambi on the banks of the Batang Hari river. According to inscriptions found on thin gold plaques (Budi Otomo 2012: 53), construction began in the 8th or 9th century. Gumpung Temple was built of brick on a square plan measuring 17.9 metres per side, and it was ringed by a court measuring 150 metres on each side. The statue is not complete but it is clear that it is of the goddess Prajñaparamita in the *vajraparyangka* posi-

tion (legs folded, the soles of the feet facing upwards). The hands are in the *vitarka mudra* position, which symbolises teaching and discussion. In terms of the clothing and jewellery, the statue has similarities with another of the same goddess found in the temple at Singosari in East Java that dates from the 13th century (Suleiman 1981: 50; Supriyatun 1986: 420). It is very probable that the Gumpung statue dates from the same century, a period characterised by political, religious and cultural contacts between Sumatra and East Java that continued into the early decades of the following century (Reichle 2007: 67). Prajñaparamita is the companion of the Adi-Buddha ('primordial Buddha'), which is why she is the mother of all the Buddhas and the source of all things material and spiritual (Hadiwijono 1971: 90). Her name means 'perfection of (transcendent) wisdom'. In this guise, she is the personification of a body of sutras in Mahayana Buddhism, and thus she enjoys great popularity. She was worshipped to grant the worshipper wisdom and learning (Hardiati 2002: 140).

<< 25. Ganesh

9th century
Banon Temple
Jligudan (village), Borobudur (district),
Magelang (kabupaten), Central Java
Stone, 150 x 114 x 90 cm
MNI 186b/4845

In Shaivism, a branch of Hinduism whose followers chose Shiva as their god, Ganesh, Shiva's son, is the god of knowledge and wisdom and the remover of obstacles. It is for this reason that his image is found in dangerous places.
Banon Temple stands on the west bank of River Progo. Built of brick, stone was only used for thresholds and lintels, thus it is unsurprising that there is almost nothing left of it today (Degroot 2009: 330331). However, it retained five superb statues, of Shiva, Vishnu, Brahma, Agastya and Ganesh, with the last being in

perfect condition. The statues and the location of the building suggest that it was a royal temple built in the 9th century.
As in all representations of Ganesh in Java, the god is shown seated in *utkutikasana* (legs horizontal and the soles of the feet touching) on two rows of lotus petals (*padma*) (Reichle 2007: 197). As the son of Shiva, he has certain attributes of his father: an aureole to symbolise his divinity, a crescent moon in his headdress, and a sacred thread (*upavita*) in the form of a serpent that traverses his chest. His trunk unceasingly draws up the delights from an inexhaustible bowl – for some the symbol of an endless thirst for knowledge – that he holds in his left fore-hand, while the others hold respectively an axe (his other left hand), a broken tusk, and beads (right).

> 26. Seated Tara

9th century
Bumi Ayu, Brebes, Central Java
Bronze, gold, 16 x 11.5 cm
MNI 6590

'Tara' means 'star'. The goddess of this name is of Indian origin and would help sailors from getting lost at sea. It was only from the 3rd century CE that she joined the pantheon of Mahayana Buddhism, in which she is supposed to have been born from the tears of compassion of Avalok-iteshvara. In other words, she is the embodiment of compassion. Each of her avatars is named after a colour: for example, Green Tara is the goddess of motherhood, and White Tara is the goddess of action. In both cases, she is shown seated cross-legged on a lotus leaf, as here. In this sculpture, the position of her right hand symbol-ises charity, and that of her left hand discord. Her headdress is adorned with a small stupa. A dot on her forehead and lower lip are gilded. A 9th-century Sanskrit inscription on the base is written in *nagari* script.

50

< 27. Manjushri

Late 9th or early 10th century
Ngemplak, near Semarang, Central Java
Silver, 29 x 16 x 16 cm
MNI 5899

Discovered in October 1927 by a farmer in the village of Ngemplak (Semarang), this sculpture can be considered intact, in spite of the lack of a base. To judge from the hair and jewels, this is the bodhisattva Manjushri kumarabhuta (Manjushri the Youth). He is seated in the *lalitasana* position, with his left leg bent horizontally, his right leg hanging down. The bulge beneath the sole of his right foot may well have been used to fix the statue in a plinth that was probably in the form of a lotus flower (*padma*). The figure's head of hair is divided in three sections. The earrings are in the form of a dharma wheel. One of the two necklaces bears an amulet that holds, beneath a rectangular structure, two tiger's teeth supposed to protect their wearer (Fontein 1990: 194). The right hand with upward-facing palm symbolises charity (*wara mudra*). The half-open blue lotus flower (*utpala*) beneath a book (*pustaka*) in his left hand represents great wisdom. Note that the palm of each hand is tattooed with a dharma wheel.

Manjushri's headdress, necklace and body were influenced by the Pala style that was introduced from India. The statue, however, was not made in India but Java, by an artist influenced by the Indian style. In fact it is possible to distinguish characteristics that are typically Javanese, such as the jewels on the hair clip, and the oval ornaments studded with precious stones that form a flower on the sacred thread, belt and upper part of the right arm.

Silver statuettes dating from the Hindu-Buddhist period, found on Javanese territory, are extremely rare. What makes this one particularly precious is the fact that it weighs no less than 8 kilos and that the silver has a purity of 92%. It was made using the lost-wax technique, which had been known on the island since before the arrival of the influence of Indian culture. This technique consists of three stages: (1) the creation of a model figure in wax; (2) the wax figure is covered with clay, then placed in the kiln to bake; the wax melts and flows out through a small hole; (3) the resulting mould is filled with molten silver and left to cool. The final statue is discovered when the clay mould is broken.

∧ 28. Inscription from Canggu

7 July 1358 CE
Pelem (hamlet), Temon (village),
Trowulan (district), Mojokerto (kabupaten),
East Java
Bronze, 36.5 x 10.5 cm

This plaque is the fifth in a set of which it is thought there were more than ten. It was issued in the year 1280 of the Saka era (1358 CE), during the reign of Hayam Wuruk, king of Majapahit, for whom this 'charter of 1358' was one of the means used to break the relations between professional goods transporters and the powerful rural nobility. Through this charter, the convoys of barges between Canggu and Terung (which were previously dependent on a local committee of elders) were placed under the authority of the court, and the taxes paid to the nobility were replaced by a contribution (*pamuja*) to religious ceremonies held at court. It was in this manner that commercial groups began to be constituted within largely agricultural communities. Place-names cited in the inscription include the village of Temon where the plaque was discovered.

> 29. Inscription from Lobu Tua

1088 CE
Lobu Tua, near Barus, North Sumatra
Stone, 92 x 30 cm
MNI D 42

The inscription is on an irregular hexagonal stone post. When discovered, the stone was broken in several fragments, some of which are still missing. The 26-line inscription was engraved in the stone in the Tamul language and script, and gives the date and reason for the erection of the post: in the month of *Masi* in the year 1010 *Saka* (February-March 1088 CE), the members of the 'Five Hundred' trade association met in Barus, where they took decisions regarding a 'captain' and a merchant. They resolved that there would be a tax to be paid in gold and exhorted everyone to maintain a benevolent attitude henceforth.

>> 30–31. *Makara*

Dieng, Central Java
Stone, 40 x 41 x 19 cm
Stone, 46 x 42 x 20 cm
MNI 410 & MNI 411

These two *makara* were found in Dieng, in northern Central Java, at an altitude of 2000 metres. The temples in the Dieng complex, eight of which have been restored, are small in size. According to an inscription made in the year 731 *Saka* (809 CE), their construction began at the latest in the 8th century. Indeed, their style matches that of the ancient kingdom of Mataram (8th–11th centuries). The Sanskrit world *makara* means 'sea monster', a mythical sea creature with a trunk that is as present in Hinduism as it is in Buddhism. Its images spread across South-East Asia from the Indian subcontinent. *Makara* are seen in temples at the bottom of stairways, beside an entrance, in niches and as gargoyles. They were also used as forms of decoration on all kinds of objects.
The Dieng *makara* stood on either side of the temple entrance. Holding a parrot in their mouths, they boast fangs, thick lips like those of a hippopotamus, a fish's tail and the trunk of an elephant, creatures that linked the *makara* to the gods: the parrot to Kama, the elephant to Ganesh, the fish to Vishnu, and so on (Wardhani 2015). *Makara* were supposed to protect the temple and ward off danger. As aquatic animals, they were associated with fertility and purification.

<< 32. Durga Mahishasura Mardini

9th century
Near Borobodur, Central Java
Stone, 130 x 60 x 35 cm
MNI 127

In Java, Durga was a very popular divinity. Study of the Sanskrit terms at the origin of her name reveals that Durga means 'the invincible killer of the buffalo demon'. It is therefore not surprising that she is represented standing on a buffalo and holding a set of weapons, including two dharma wheels, a sword, a spear and a shield. In addition to other adornments, she wears a jewellery snake coiled on her hips.

It is said that by practising asceticism and meditation, Mahishasura, the buffalo-demon, obtained from Brahma the promise that he would not die by the hand of a man. He thus believed himself immortal, as he never imagined that a woman might have the strength to kill him. Consequently, he threw himself into the endless battle between the gods and the demons, even defeating Indra, the supreme god, prompting Shiva and Vishnu's anger. The pair's faces produced a blinding light which, combined with that from the faces of the other gods, formed a mountain of light that suddenly turned into an extremely beautiful woman: Durga. All the gods gave her weapons, whereupon she engaged in a ruthless battle against Mahishasura. In spite of indulging in successive metamorphoses (buffalo, lion, elephant, then a buffalo again), the demon was unable to overcome Durga, and it was during this last apparition that she slaughtered him.

<< 33. Seated Maitreya

Late 9th–early 10th century
Komering (river), Palembang, South Sumatra
Bronze, 24.5 x 13 x 8.6 cm
MNI 6025

This seated Maitreya was found in 1929 with two other Buddhist statues (a standing Buddha and a Bodhisattva Avalokiteshvara) near Palembang, at the bottom of the River Komering (Kempers 1959: 174–176). Maitreya means 'Love' and he is the Buddha who will bring happiness, joy and hope to humanity in the future. His attribute is the small stupa above the *padma* (lotus) adorning his crown. The proportions of the figure and the fineness of the jewellery suggest this statue was influenced stylistically by the art of the Shailendra dynasty (Java, 8th–9th centuries). Artefacts of this type have been mostly found around Palembang and in the central part of Sumatra (Suleiman 1981: 54). This is not surprising as, by the middle of the 9th century, the last representative of the dynasty fled to Palembang, his mother's hometown, where he became king of Sriwijaya. Sumatran sculpture gradually forged its own identity. One of the distinctive features of these Buddhist works is the style of the hair and the arrangement of Maitreya's has affinities with representations of Avalokiteshvara in the Malay Peninsula (Suleiman 1983: 331). Throughout its golden age, as from the 8th century, the kingdom of Sriwijaya controlled ports on both sides of the Straits of Malacca, thus on the island of Sumatra and the peninsula. It is therefore very possible that some sculptures were brought to Sumatra, but just as likely that Sumatran sculptors went to work in Malacca. Whatever the facts of the matter, both the style and place of creation of this statue of Maitreya suggest that it can be dated to the golden age of the kingdom of Sriwijaya.

> 34. Inscription on a sacrificial post (*yupa*)

4th century CE
Muara Kaman, Kutai, East Kalimantan
Stone, 169 x 39 x 29 cm
MNI D 2a

This inscription in the Sanskrit language, written using Pallava script, gives the names, on the occasion of a ceremony, of Mulavarman, 'lord of kings', his father Asvavarman, 'founder of a noble race', and his grandfather, 'the great Kundunja, lord of men'. It specifies that the post had been erected by the head Brahman, seemingly in thanks for religious services rendered to Mulavarman, whose offering was impressive. It should be noted that the grandfather has an indigenous (Austronesian) name, whereas those of his son and grandson are of Indian origin. According to the inscription, Mulavarman's religion was Vedism. Thus, this post is a vestige of the first introduction of Indian culture to the island of Borneo.

‹ 35. *Nandi*

13th–14th century
Padang-Lawas, North Sumatra
Stone, 40 x 54 x 39 cm
MNI 326b/3290

The *nandi* is the mount ridden by
Shiva, the supreme god of Hinduism.
The creature's extraordinary eyelash-
es and neck adorned with bells make
this an outstanding representation.
It is very rare to find Hindu sculptures on Sumatra, and all the more
so due to the fact that during the peak
splendour of the Sriwijaya kingdom
(around the year 1000), Padang Lawas
was an important centre of Buddhist
culture. The style of the sculpture
suggests that it dates from the golden
age of the kingdom of Majapahit
(13th–14th century) and thus that it
was brought to Padang Lawas by
Hindus.

ˇ 36. Inscription
from Kedukan Bukit

682 CE
Kedukan, Palembang, South Sumatra
Stone, 27 x 49 x 34 cm
MNI D 146

This irregularly shaped stone has
certain pieces missing and some of the
characters can no longer be identified.
This inscription, written in Old Malay
and using the Pallava alphabet, nar-
rates the sea journey (*siddhayatra*) of
a king who set out on 23 April 682 CE
('the eleventh day of the half-moon
of the month of *waisaka* in the year 605
of the Saka calendar'). On 19 May of
the same year, he left a delta with
20,000 soldiers and 200 chests of
provisions on board, while 1312 men
travelled by land. The joy and relief
was great when, on 16 June, they all
arrived in the kingdom of Srivijaya,
an event that marked the start of a
period of prosperity.

63

< 37–39. Brahma, Vishnu and Shiva (Trimurti)

14th–15th century
Air Bersih, Palembang, South Sumatra
Brahma (MNI 6033): Bronze, 55 x 14 x 17.5 cm
Vishnu (MNI 6032): Bronze, 57 x 28 x 22 cm
Shiva (MNI 6034): Bronze, 50.7 x 16 x 24 cm

Brahma, Vishnu and Shiva are the three main divinities in the Hindu religion and are known together as the *Trimurti*. Brahma is the Creator and an incarnation of the soul of the world. He exists as an autonomous entity, without beginning or end. Vishnu is the protector of the universe and humanity and has ten different avatars. Shiva is the god of destruction and is manifested in different forms depending on his many functions. The Canggal inscription (723 CE) informs us that, in Indonesia, Shiva was considered superior to Brahma and Vishnu. Each of these gods is shown standing on his mount: Brahma on a goose, Vishnu on the *garuda* (a mythical creature) and Shiva on his bull (*nandi*). Sculptures of the gods on their mounts are rare. These are fine examples of the style of the Majapahit kingdom, which had continuous relations with Sumatra in the 14th century.

>> 40. Standing Parvati

14th or 15th century
Probably East Java
Stone, 62 x 30 x 24 cm
MNI 114 c/3546

The dating of this statue is based on the style of the clothing and jewels worn by the figure whose iconography reveals must be Parvati, the wife of Shiva. She is shown standing serenely on two rows of lotus petals. Two of her four hands are placed on her diaphragm, while her two others hold a fly-whisk and a string of beads. It is thought that this representation is that of a dead queen as, in East Java, it was the custom to represent the dead twelve years after their death as a god or goddess they had worshipped during their lifetime. The statue has only two arms, which are shown in the meditation position. If the statue

1149.

had had four, two would have been in the same position of meditation, while the other two would have borne attributes of the divinity. Parvati's eyes are either closed or half-closed.

<< 41. Fountain (*gargoyle*)

13th–14th century
Sirah Kencong (village), Wlingi (district),
Blitar (kabupaten), East Java
Stone, 105 x 75.8 cm
MNI 383a/4385

The holes in the top and bottom of the sculpture, which are connected, suggest that this artefact was a gargoyle or fountain. All the elements represented have a link with the fairly complex history of the churning of the ocean of milk, the *samudra manthana*, accounts of which in the ancient texts of different traditions do not necessarily conform. In Hindu mythology, the ocean of milk was the topmost layer of a series of concentric oceans. The underlying premise of the story is the quest for the nectar of immortality through the battle between good and evil. As fervently as the gods and demons wished to take possession of the nectar, they were unable to as it was hidden in the ocean of milk. In order for it to appear, the two sides had to unite their forces for an eternity to churn the primordial liquid and overcome a series of obstacles.

^ 42–44. Heads of figurines

14th–15th century
Trowulan, Mojokerto, East Java
MNI 8653b: Earthenware, 9 x 4.9 cm
MNI 8656a: Earthenware, 6 cm high
MNI 8722: Earthenware, 6.8 cm high

Known in Indonesia since the Neolithic, earthenware experienced great development and became more complex in the kingdom of Majapahit (1293–c. 1500). Many figurines have been found in Trowulan, the kingdom's first capital. Some were whole but most are statuettes that are missing their head, or heads that are missing their body. The faces are of locals or foreigners: Chinese, Tartars, Indians, Persians, Thais, Vietnamese and Europeans (Portuguese and Greeks), which together provide evidence of maritime trade with a number of countries. The function of these figurines is unknown; were they toys, puppets, moneyboxes, ritual objects or simply decorative?

<< 45. Amulet holder

9th–10th century
Plosokuning (hamlet), Wonoboyo (village),
Klaten (kabupaten), Central Java
Gold, 11 x 2.5 cm; length of the strap: 60 cm
MNI 8916

This amulet bag was part of the Wonoboyo Hoard, discovered by agricultural labourers in 1990 in an earthenware jar 2.5 metres beneath ground level. The treasure, which weighs 16.9 kg in all, consisted of various objects made of gold (14.9 kg) and silver (2 kg). It dates from the reign of king Balitung (899–910 CE) of Mataram. Rectangular and made completely of gold, this small flapped bag is attached to a long, magnificently worked strap of gold thread that passes through two holes on the sides of the bag. The entire surface is decorated with repoussé images of vine leaves and flowers, which are enhanced at the bottom by chasing and engraving. In addition to these basic motifs, we also see images of *sangkha* (conches) and *chakra* (discs). For this reason, it is thought that the bag may have held sacred objects or incantations for disciples of Vishnuism.

<< 46. Shiva Mahadeva

9th century
Gemuruh (hamlet), Banyukembar (village),
Leksono (district), Wonosobo, Central Java
Gold, 20.5 x 11 x 3.5 cm
MNI A 24/517b/4565

This gold plaque was embossed and then finely chased and engraved. The relief shows Shiva Mahadeva standing on a lotus flower. His right fore hand holds a trident and the other right hand a string of beads; the left fore hand holds a jug and the other a fly-whisk. Another characteristic of Shiva is the third eye in the middle of his forehead. His sacred thread is in the form of a serpent whose head rests on his left shoulder. The cloth between his waist and hips is decorated with regular, dotted lines and is attached by a sash whose ends hang over his thighs. Furthermore, over his hips Shiva wears a tiger skin whose

head rests on his right thigh. Behind his head is an aureole and the rounded top of the plaque serves as a halo (*prabhamandala*). The inscriptions in Old Javanese to the right of the trident probably give the name of the craftsman (NBG 1903: 118) and a spell (Martodikromo 1999: 44).

> 47. Necklace

9th century
Plosokuning (hamlet), Wonoboyo (village),
Klaten (kabupaten), Central Java
Gold, 25.5 x 1.5 cm
MNI 8865

Discovered in 1990, this necklace was an element in the Wonoboyo Hoard that evinces the agricultural and maritime wealth of the kingdom of Mataram (8th-10th century). Like many artists in South-East Asia, India and China, those in Java took inspiration from their natural environment. This necklace is an example with the ornamentation that imitates the shells of freshwater snails that are only found on their island: *Sulcospira testudinaria*. It was made using the lost-wax technique (see cat. 27) and by fusing the gold shells to small tubes made from the same metal.

>> 48. Scene from the *Ramayana*

Kediri, East Java
Gold, 14 x 12 x 1.5 cm
MNI A 110/6914

This embossed gold plaque describes an episode from the epic tale, the *Ramayana*. Led by Hanuman, the army of monkeys built a bridge over the ocean to connect the southernmost tip of the Indian subcontinent with the kingdom of the demon Ravana on the island of Lanka, where Ravana holds Sita, Rama's wife, whom he has abducted. The plaque's artist succeeded in making this episode humorous by inserting representations of crabs with threatening claws in the ocean, whose bites the monkeys avoid by lifting their feet into the air. The crabs take Ravana's side and attempt to impede the construction of the bridge, whereas the monkeys help Rama to free Sita (Fontein 1990: 114). At the top of the plaque, the square holes seem to represent the bridge under construction. Similar representations are seen in the sculptures in temples in Prambanan (Central Java) and Panataran (East Java).

>> 49. Mirror (*darpana*)

14th–15th century CE
Krai (hamlet), Semanten (village),
Pacitan (kabupaten), East Java
Bronze, 26.8 x 15.8 x 0.5 cm
NMI 1108 f/ 3872

During Hindu ceremonies, mirrors were sometimes used during the purification of statues of gods. On this occasion, it was not the statue – whether of clay, gold or silver – which was being washed and sanctified but the mirror that reflected its image. This round mirror with a smooth, undecorated front face has a handle in the form of an inverted T. On the back, the small raised centre point surrounded by concentric circles is suggestive of a nipple.

< 50. Standing Buddha

8th–9th century
Mouth of the River Komering,
Palembang, South Sumatra
Bronze, 40 x 12 x 8.8 cm
MNI 6023

Buddha is seen in the *samapada* dance position with his feet side-by-side on the ground. His right hand makes the sign of *abhaya* ('absence of fear'), while his left hand holds the draped garment that covers his shoulders. This last stylistic element, the form of the necklace taken by the pleats of the garment around his neck, the swallowtail train, and the bulge on the figure's head attest the influence of the art of Pala India. The protuberances beneath the feet were used to hold the statue on a plinth.

> 51. Ganesh

9th–10th century
Sumatra (Padang Lawas?)
Bronze, H: 20 cm
MNI 534 a/4839

To judge by the figure's style and
headdress, the statue was made in
Central Java in the 9th or 10th century.
Unlike most representations of
Ganesh, the bowl of delights is in his
left hand while his right hand holds
a broken tusk.

77

< 52. Avalokiteshvara

9th–10th century
Ledok, Bagelen, Central Java
Bronze, 32 x 16 x 11 cm
MNI C 119/614

Avalokiteshvara is one of the bodhis-attvas that guard the universe between the reign of Buddha Gautama and the reign of the Buddha of the future, Maitreya. Avalokiteshvara refused to enter nirvana as long as mankind needed him.

> 53–54. Avalokiteshvara

10th–11th century
Rantaukapastuo, Muara Tembesi (district),
Batanghari (kabupaten), Jambi, Sumatra
Bronze gilt
Musée provincial de Jambi, 04.093 & 04.094

These two representations of Avalokiteshvara were found by chance
in 1991. East Javan style is combined
with influences from South India.
The statuettes were probably brought
to Sumatra by merchants or Buddhist
priests who stopped at the island
during a port of call.

79

∧ **55. Stele from Kubu Kubu**

905 CE
Unknown source
Bronze, 35.5 x 6 cm
MNI E. 75

The inscription in Old Javanese script and language tells us that king Balitung (899–910 CE) of Mataram gave the village of Kubu Kubu to Rakryan Hujung Dyah Mangarak and Rakryan Matuha Dyah Majawuntan as a reward for their conquest of Bantan. Bantan may signify Bali, as *bantan* means 'sacrifice', while *bali* means 'offering'.

> **56. Agastya**

8th–9th century
Semarang, Central Java
Stone, 40 x 35 x 90 cm
MNI 61

Agastya is Shiva's leading disciple. He is an old, bearded man with a large belly and wears a sort of crown on his head. In Indonesia, he is considered one of Shiva's avatars.

> **57. Inscription from Kota Kapur**

686 CE
Penagan, Bangka, Bangka Belitung
Stone, 166 x 31 x 25 cm
MNI D. 90

This inscription in Old Malay and Pallava script commemorates the conquest of the island of Bangka by Sriwijaya, and contains a warning that those who do not submit will be cursed, while those who do will prosper. The conquest of West Java is stated at the end of the text.

∧ 58. Fragment of a figurine

14th–15th century
East Java
Clay, 6.5 x 4 cm
MNI 8721

This fragment shows the headdress of a woman from Trowulan, a town in the kingdom of Majapahit.

> 59. Fragment of a figurine

14th–15th century
East Java
Clay, 6 cm
MNI 8763

The figurine probably represents an ascetic or a religious figure.

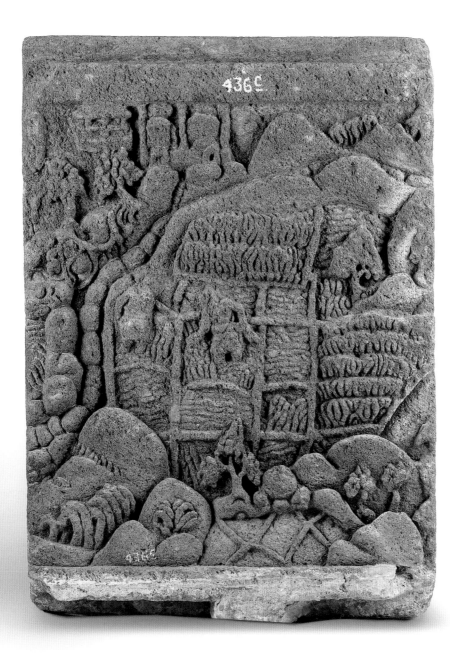

∧ 60. Relief

14th–15th century
East Java
Stone, 8.5 x 43 x 30 cm
MNI 436c

The relief depicts the rice paddies
of Java in the 14th or 15th century.
We see canals, raised pathways, and,
in the distance, mountains and rivers.
The irrigation of the paddies was
directed by an official called the
huluair.

83

^ 61. Mirror handle

14th–15th century
Kallak, Pringkuku, Madiun, East Java
Silver
13.7 x 14.4 x 2.3 cm
MNI 1178

Mirrors were not only used for
practical purposes: they also had
the power to protect against magic.

> 62. Plate

900–950
Central Java
Silver
2.5 x 19 cm
MNI 1737

This plate was probably used to
hold religious offerings. The figure
at the centre is Sita, one of the
characters in the *Ramayana*.

^ 63. Plate

8th–9th century
Origin unknown
Silver, 3.5 x 25 cm
MNI 1736/A 309

Dating from the period of the Song
dynasty in China, this plate is finely
decorated. Two fish in the centre
have an elephant's trunk.

85

^ 64. Bowl

9th–10th century
Surakarta, Central Java
Gold, 5.5 x 10 x 10 cm
MNI 1715 c/3392/A128

The form of this bowl was probably an imitation by Javanese goldsmiths of ceramic bowls produced in Tang China (618–907).

˅ 65. Stamp ring

9th–10th century
Found off the north coast of Jakarta
Gold, 2.7 cm
MNI 18900

The ring was found in the cargo of the Intan shipwreck. The Sanskrit inscription means 'burning'.

66–69. Treasure from the Intan shipwreck

These objects were found in the cargo of a shipwreck off the coast of Jakarta, at the south-west tip of the Seribu Islands, close to the drilling platform of the Intan oil company. The finds were brought to the surface and investigated in the 1990s. The boat was carrying several thousands of Chinese and Thai ceramics made between the 10th and 13th centuries, plus the *badong* (see cat. 74) and gold rings. The boat was probably sailing from the Sriwijaya capital, Palembang, to Central or East Java.

˅ 66. Ring

9th–10th century
Found off the north coast of Jakarta
Gold, 3 cm
MNI 18899

This ring is decorated with rinceaux on a double lotus-flower petal.

˅ 67. *Masa*

9th–10th century
Itvar cargo, Lansui, north coast of Jakarta
Gold, 1.2–1.5 x 1.2–1.3 cm
MNI 18893-96

The term *masa* comes from India. It refers to a commercial coin used during the Javanese era. One *masa* weighs 2.4 g. *Ma*, an abbreviation of *masa*, is written on the convex face, while a lotus flower is shown in a rectangular frame on the other side.

˅ 68. Chastity cover (*badong*)

9th–10th century
Found off the north coast of Jakarta
Gold, 7 x 6 cm
MNI 18897

Chastity covers like these were worn outside clothing by high-born or noble women when their husband was away, but sometimes also by ascetic women to announce their celibacy.

˅ 69. Lotus-shaped container

875–975
Wonoboyo, Klaten (kabupaten), Central Java
Gold, 10 x 9 cm
MNI 8930

The religious significance of the lotus is derived from its manner of growth: its roots in the mud (the earth), the stem in the water, and its leaves and flowers in the air.

86

> 70. Panel

16th century
Mantingan mosque, Japara (kabupaten),
Central Java
Stone, 38 x 59 x 11 cm
Ranggawarsita, Semarang Provincial
Museum, Central Java, 04.597

This stone panel has two faces. On
the front we see a stylised elephant
among lotus plants that cover the
panel's entire surface. The other side
shows a fisherman and, behind him,
two larger warriors, one of which
holds a bow. Although their faces are
damaged, it is possible to make out
that one of the warriors is Rama and
the other his brother Laksmana in
a scene from the *Ramayana*. The
sculpture is influenced by Chinese art,
but this is not surprising as it
was carved by a Chinese Muslim.

> 73. Incense burner

1st–2nd century (Han dynasty, China)
Jambi, Sumatra
Earthenware, 21.5 x 18.5 cm
NMI 2708

The holes in the lid of this artefact are clear evidence that it was used for burning incense. Its conical shape and relief decoration on the lid conjure up the idea of a mountain. Called *bo shan lu* (mountain peak) in Chinese, the burners were associated with the Taoist desire for purity and immortality. Ceramics produced during the Han dynasty (206 BCE–220 CE) have similar forms and the robustness of the bronze jars of the earlier Zhou period (480–221 BCE). Despite the article remaining underground for a long time, the green of the enamel caused by lead in the glaze is covered with silvery highlights. Given the toxic nature of the enamel, it is thought that these objects were not used to hold foodstuffs but had a ritual and/or funerary function. Han ceramics found in Indonesia – always in the west of the country – were above all used in ancestor worship. It is believed that such objects were brought back to Indonesia from abroad by their owner and that they became family heirlooms. Also that they became buried by natural catastrophes or in a funerary context.

∧ 71. Panel

16th century
Mantingan mosque, Japara (kabupaten),
Central Java
Stone, 38 x 36 x 11 cm
Ranggawarsita, Semarang Provincial
Museum, Central Java, 04.596

Similar to the previous panel (see cat. 70), here the front is carved with rinceaux, while the other side also depicts a scene from the *Ramayana*, in which three monkeys help Rama search for his wife.

>> 72. Cargo from a shipwreck off Buaya Island

In 1989, the wreck of an ancient ship was discovered off Buaya Island, to the south of Singapore. Its cargo, containing 30,000 wares of porcelain produced in the kilns of Guandong and Fujia province in southern China, was brought to the surface. Most of the objects date from the Song dynasty (960–1279). The ship was probably sailing to Sumatra or Java where its wares would have been sold.

‹ 74. Dish

14th century (Yuan dynasty, China)
Halmahera, Maluku
'Blue and white' porcelain, 8 x 46.5 cm
NMI 1134

Made in Jingdezhen in Jiangxi province, southern China, this superb dish attests a mix of influences: the rim is reminiscent of Middle Eastern metal objects and the decoration of Persian ceramics, while the wave motif is typical of the style of the Yuan dynasty (1279–1368). Usually, artefacts like this one have a dark blue decoration on a white ground but here the reverse is true. The dish was found on Halmahera in the Maluku Islands, renowned for their spice production. It was as a result of their prosperity derived from spice production that the islands' inhabitants could afford luxury objects that also served as heirlooms and status symbols. Made for export to the Middle East, where they were used in communal meals typical of Islamic tradition, these dishes were also distributed in Indonesia.

› 75. Large vessel

Between 1572 and 1620
East Kalimantan
Porcelain ware from Jingdezhen,
50 x 56.5 cm
NMI 3853

Beneath the flared rim of this lidless and rather thick porcelain container are six Chinese characters that indicate it was made in the reign of the Wanli emperor (1572–1620 CE) during the Ming dynasty. The blue underglaze decoration on the outer walls shows two five-clawed dragons and a lotus flower. The vessel was used as an urn or a lamp of 'eternal fire' (so-called because of the amount of oil that it could hold), probably in the temple where it was found. Such objects are rare due to the difficulty of their manufacture.

› 76. Jar

16th century
Serang, Banten, Java
Porcelain ware from Jingdezhen, 36,2 x 39,9 cm
NMI 1745

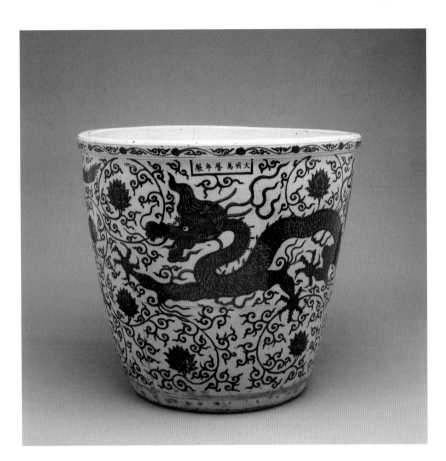

This thick porcelain vessel and the wine it would have contained were imported from China. The greyish-blue underglaze decoration is of refined execution. The central motif – the eight immortals of Taoism – is accompanied by the representation of one of the three Chinese 'gods of good fortune', Shou Lao, the god of longevity, and his companions, the roe deer that scour the mountain in search of the mushroom of immortality. The theme of good fortune and long life is also signified by the clouds in the background, a favourite motif of the Ming (15th-16th centuries), which the Javanese employed in several artistic disciplines. Lastly, the cypresses symbolise long life and endless friendship.

ang in China. Given that even the underside of this ewer is covered in enamel, the process can only have been carried out by immersing the object in the liquid. Its rigid form is a reminder that an attempt was originally made to copy the angular form of the bronze ewers of the earlier period (see the straight edges on the upper, dish-shaped part). Whereas the spout is in the form of a chicken, the curved handle takes the form of a dragon. Initiated during the Eastern Han dynasty (25-220 CE), green porcelain was produced until the early decades of the Song dynasty

(960-1279). It was under the Song and earlier Tang dynasty (618-907) that the quality of production increased considerably, with the stoneware becoming thinner and smoother, and the decoration more opulent. These objects were exported to Japan, South-East Asia, India, Persia and Egypt. In Bali, Sumatra and above all around the temples of Java, both fragmentary and intact ewers have been found. Stoneware artefacts like this one were heirlooms though they had probably originally been imported as domestic objects (Adhyatman 1982; Flines 1969; Ridho 1979; Vainker 1997).

>> 77. Amphora

7th century (Tang dynasty, China)
Palembang, South Sumatra
White porcelain stoneware, 37.5 x 19.4 cm
NMI 3068

In the shape of a Hellenistic amphora, this vessel has a relief decoration of flowers and masks suggestive of Central Asian metalworking, and handles in the form of Chinese dragons. It dates from the Tang dynasty (618-906), during which time China grew into a sizeable commercial power. The capital, Xi'an, at the eastern end of the Silk Roads, was then one of the world's most cosmopolitan cities. The presence of this vase in Sumatra is probably explained by the strong trading relations between China and the Sriwijaya kingdom, of which Palembang was the capital from the late 7th to early 11th century.

> 78. Ewer

4th–5th century (Eastern Jin dynasty, China)
Bengkulu, Sumatra
Stoneware, 29 x 21.3 cm
NMI 1728

Often called 'green porcelain' due to the colour of its enamel (as a result of the use of iron oxide), this type of stoneware was made in the kilns at Yue in the coastal province of Zheji-

∧ 79–81. Treasure from the Belitung shipwreck, also called the Batu Hitam shipwreck

This wreck, discovered in 1998 at a depth of 18 metres in the Gelasa Strait by fishermen from Batu Hitam, was the first Arabian or Persian dhow to be found in Indonesian waters. It is thought the dhow sank around 830 CE after hitting a rock. It appears that at the time direct commercial relations already existed between China and the Arabian Peninsula. A great part of the cargo was made up of Chinese wares from the end of the Tang dynasty (9th century), among which 60,000 ceramic bowls made in Changsha in the province of Hunan. The goods were protected in large storage jars specially produced for exportation purposes. These type of ships left South China, stopped in Central Java, then passed through the Sunda Strait before heading for West Asia or the Middle East.

∧ 82. Dish

12th–14th century (Song
and Yuan dynasties, China)
Jambi, Sumatra
Stoneware, 7.5 x 35.5 cm
MNI 935

This dish is a celadon ware decorated
inside with an independently moulded
motif of a four-clawed dragon (the
dragon of the nobles) on the bottom,
and with rinceaux on the walls. It was
manufactured in Longquan in the
province of Zhejiang. The wares from
this city are some of the best in the
world.

97

< 83. Jar

15th century
Palopo, South Sulawesi
Porcelain
MNI 923

The thick porcelain walls of this jar are cream in colour with a blue decoration. The primary motif is the mythical creature that symbolises kindness and good fortune – the *qilin* – with the *fenghuang*, one of the four Celestial Animals of Chinese mythology. It is also found in Indonesia: some of the elements were taken up in the *buraq* motif of Cirebon, and the decoration of the *tok wi* in the catalogue (cat. 136) includes a *qilin* that may have been the work of a Chinese immigrant (Adhyatman 1982, de Flines 1969, Ridho 1979, Vainker 1977). The jar was made in Vietnam by potters who adopted the 'blue and white' technique invented by the Chinese. The trade route for jars of this kind left Vietnam, and followed the coasts of the Malay Peninsula, Sumatra and Java, before heading to South Sulawesi, the centre of the spice trade.

> 84. Ewer

9th century (Tang dynasty, China)
Stoneware, 35.5 x 8.9 cm
MNI 2941

This jar-shaped container held drinking water. The yellowish enamel and motifs (dragon, lizards) are typical of the Tang dynasty style and kilns at Changsa (Hunan province). Comparable pieces have been found in Japan, Iran and Pakistan.

< 85. Dish

17th century (late Ming dynasty, China)
Lampung, Sumatra
Porcelain, 8 x 36.5 cm
NMI 1495

The polychrome enamel of this dish has been given a decoration in Arabic script. The underside of the rim has a few impurities. The dish is an example of what is called Swatow or Zhangzhou ware (a city in Fujian province, China). The texts are taken from the Quran and the sayings of the Four Companions of the Prophet. Islam and Arabic script spread in China from the time of the arrival of the first Islamic merchants in the 8th century. From the reign of the Zhengde emperor (1505–1521) of the Ming dynasty, many court officials were Muslim, who commissioned 'blue and white' ware with Arabic inscriptions for use in offices. Some potters learned Arabic, with the result that inscriptions can be understood without difficulty, but others demonstrated more goodwill than talent. This dish was probably an article in an order for the markets of the Middle East or South-East Asia (de Flines 1969; Li Zhiyan 1989; Ridho 1979; Vainker 1977).

˅ 86. Seal

1879
Sultanate of Aceh
Gold, 1.5 x 21.5 cm
MNI 13209/E.154

This seal bears witness to the importance of the sultanate of Aceh, which controlled trade in the archipelago in the early 17th century. Although its form resembles that of the stamp used by the Mughal emperor Jahangir as from 1619, the décor is similar to that of Ming wares of the 16th and 17th centuries embellished with Arabic calligraphy. The script used here is *naskh*. Dated 1296 in the Hijri calendar

(1879 CE), the seal is engraved with the name of the reigning sultan (Alaiddin Muhammad Daud) in the large central circle, and the names of his eight predecessors in the smaller surrounding circles. It is this layout that gives this type of seal its name: *sikureueng*, meaning 'new' in the language of Aceh.

< 87. Crown (*ketopong*)

1845–1899, acquisition after 1967
Kutai, East Kalimantan
Gold, diamonds, cloth, 22 x 31 x 20 cm
MNI E1329

An article in the regalia of the sultanate of Kutai Kartanegara, this *ketopong* made in the 19th century was worn by the king during his coronation and other important ceremonies. We know that it was worn by Sultan Muhammad Sulaiman (1845–1899). It is an outstanding example of the goldsmith's art that demonstrates the expertise of the local craftsmen. Carl Brock (*De Koutei à Bandjirmasin. Voyage à travers Bornéo*, 1890) states that the sultan had six to eight goldsmiths working for him.
The crown weighs about two kilos. Its form is referred to as *brunjungan* ('high and round') and is distinguished by the presence on the front of a seven-storey pagoda decorated with flowers and garlands. Beneath, the crown is engraved with seven diamond-studded flowers, and behind by an image of Garuda. The upper section is adorned with a tuft of filigree leaves and flowers. The crown was used during the coronation of the most recent sultan, Haji Aji Muhammad Salehuddin II on 22 September 2001.

> 88. *Besurek* cloth

Palembang, South Sumatra
Cotton, 225 x 89 cm
MNI 28191

This cloth has been decorated with Arabic calligraphy on a dark ground. Introduced to Bengkulu in the 17th century by tradesmen and workers

from India, the motif of the *besurek* was enhanced by the local tradition and the use of Jawi calligraphy (the Arabic alphabet adapted to write Malay). In the past, this type of cloth was used above all in traditional ceremonies, but today the *besurek* has been transformed into a decorative pattern of spirals and loops that have only a remote resemblance with Arabic calligraphy. Other motifs (magnolia, jasmine, cloves, fern leaves, the tree of life) symbolise the lushness of tropical forests. Formed by an arrangement of Jawi characters, the motif of the *kuau* (a bird in the pheasant family) is a symbol of splendour, wealth and glory. And while the motif of the moon glorifies the Creator of the universe, calligraphy as such is an expression of respect for the All-Merciful.

< 89. Container (*pekinangan*)

Cirebon, West Java
Wood, 50 x 63 x 26 cm
MNI 23606

A *pekinangan* holds the basic ingredients for chewing betel (*menginang*): betel leaves (*sirih; piper betle*), arec nuts (*pinang; areca catechu*) and a little chalk (*kapur*). The flavour of the mix can be enhanced by the inclusion of gambier extract (*Uncaria gambir*) and/or tobacco. *Menginang* is a universal and traditional form of hospitality in all Indonesian milieus. On some islands, it is also offered to the spirit of the dead.

Lying on the north coast of Java, Cirebon has an artistic heritage arising from commercial relations that mix Javanese, Indian, Chinese and Arab influences, as is evidenced by this container: the mythic animal called *paksinagaliman* is a fusion of the *paksi* (eagle, of Islamic origin), the *naga* (dragon, of Chinese origin) and the *liman* (elephant, of Indian origin). The subject is the battle between good and evil in order to attain perfection. The *pekinangan* is always accompanied by other instruments: a *kacip* (betel scissors), a *tutukan* (mortar for crushing the chalk, and the betel nuts and leaves) and a *paidon* (spittoon).

The *menginang* tradition is gradually disappearing but it can still be found on some islands.

105

THE EARLY MODERN PERIOD

SEVENTEENTH AND EIGHTEENTH CENTURIES

Europeans have invested in Nusantara since the beginning of the sixteenth century, with the Portuguese the first to land, who marked their arrival with the conquest of Malacca in 1511. The city port in the Strait of Malacca was finally conquered after being besieged for 40 days by about 1,200 soldiers on 18 ships, after which the Sultan – Sultan Mahmud Shah – was expelled from Malacca.

The attack and the conquest of the port towns and the kingdoms of Nusantara were among the reasons for the Portuguese presence in the region. Generally governed by Muslim kings, these were the main areas for commercial and political activity. In addition to Malacca, a large city port to the west of Nusantara, the Portuguese also conquered Ternate, in the Moluccas, a kingdom of great importance, whose power was based on the spice trade. Other smaller kingdoms (reinos) and kadatuan of Nusantara, such as Pasei in Sumatra, and Tidore, Jailolo, and Hitu in the Moluccas, were attacked and conquered by the Portuguese, who also seized the Muslim merchant ships that they encountered on their journey from Mozambique to the Indian subcontinent.

The diminishing power of Islam and the spread of Christianity were among the reasons for the Portuguese arrival in the East. The historical roots can be attributed to the Crusades, which had been conducted in waves since the end of the eleventh century. In 1502, for example, in the spirit of the Crusades, King Manuel II sent Vasco da Gama to break the Muslim monopoly in shipping and commerce in the Indian Ocean. A similar undertaking was also conducted by the Spanish. In addition to the religious background behind this endeavour, the arrival of the Portuguese and other European countries in Nusantara was conducted with the aim of gaining direct access to spices. The highly sought-after spices were expensive commercial products in Europe at that time, which was due to the political upheaval and the wars in Western Asia and the Mediterranean. Getting direct access to the areas where the spices were grown would enable the European traders and entrepreneurs to reap enormous benefits.

The rivalry inspired by the victories of the Portuguese and other European countries in Nusantara also made them more eager to push for further conquests, which their large stockpiles of weapons and superior technology certainly made possible.

The following reason is also associated with the development of science and technology in European countries. By the middle of the fifteenth century, Europeans had begun to build large vessels with a carrying capacity of 500 to 800 people, and capable of sailing the high seas armed with 30 to 40 cannons. In addition, they also had the skills to create navigation charts, to determine latitude and longitude to aid navigation, to read and use the stars, especially those of the Southern hemisphere, to steer by means of a wheel connected to the ship's hull, and to use the magnetic needle and the compass, all tools that were very important when crossing vast oceans and visiting unknown areas. The exploration of the ocean by the Portuguese and the Spanish was then followed by other European countries, such as the Netherlands, England, and France, among others.

As mentioned above, the attacks and conquests realized by the European countries were not achieved without strong resistance from several of the kingdoms and kadatuan of Nusantara. For example, the Portuguese with their 18 ships and 1,200 sailors took around 40 days to conquer Malacca.

It is not surprising that Malacca had a powerful army at its disposal. During its heyday, Malacca was extremely dominant in the Strait of Malacca, maintaining the powerful army that it needed in order to ensure the safe passage of its shipping. The safety of the Strait of Malacca led many merchants and seafarers to take refuge in this city.

By the time the Portuguese arrived in the region, Malacca was the largest city port in Southeast Asia. Sailors and merchants from at least a dozen countries did their trading in this city port, arriving from places as far flung as Egypt to the west, China in the north, and the Moluccas to the east. That is why it was not surprising to be able to hear as many as 84 languages being spoken there. Various types of vessel were moored at the port of Malacca, including boats, junks, pangajavas, lancara, and liners, not to mention the huge variety of products that were marketed in the city. The value of the trade that took place in Malacca reached levels as high as hundreds of millions of cruzados every year.

As a political force, Malacca established relations with several kingdoms and kadatuan in Sumatra. In addition, a number of kingdoms and kadatuan were vassal regions of Malacca, such as Rokan, Rupat, Siak, Kampar, Indragiri, Tongkal, Jambi, and Palembang. Political ties were established with a number of kingdoms in Sumatra, as well as in Java. Demak was one of the kingdoms of Java that maintained close relations with Malacca. There were political links and strong emotional ties between Malacca and the kingdoms and kadatuan at the point when the Portuguese took over Malacca. Almost all of the kingdoms and kadatuan tried to help Malacca to protect the city, but unfortunately, the power of the Portuguese weaponry proved too strong.

Traces of flourishing commercial activity were also found in other city ports such as Aceh, Palembang, Banten, Cirebon, Semarang, Gresik, Surabaya, Banjarmasin, Gowa, Ternate, Tidore, and Hitu. Like Malacca, various groups of merchants and sailors also came to these city ports, and their conquest at the hands of European countries did not mean that the political and economic power of Nusantara was immediately destroyed. For a while, when European countries were first present in Nusantara, the conquest of a

kingdom or a city port led to political and economic activities emerging in other areas. The conquest of Malacca, for example, triggered the emergence of Aceh and Banten as a consequence of the expulsion of so many Muslim sailors and merchants, who took to using the alternative sea routes on which the two city ports were located. Aceh is located at the northernmost tip of the island of Sumatra, while Banten is located in the Strait of Sunda, a strait crossed by sailors and merchants from the Indian Ocean seeking to access the commercial towns of the Java Sea and eastern Indonesia. Aceh and Banten were also able to provide the various commercial amenities sought by Muslim sailors and merchants.

The European countries competed with each other in their bid to establish their influence. During the first days of their presence in Nusantara, the Portuguese and the Spaniards competed to impose their influence on the Moluccas. The Portuguese dominated the game, and the Spaniards had to leave Indonesia to colonize somewhere else, namely the Philippines. In the period that followed, the Portuguese competed with the Dutch, who eventually came out on top. In the wake of their defeat, the Portuguese concentrated on their regime in Timor.

The Dutch victory over the Portuguese (and later over the British) was due to several factors. First and foremost was the organizational capacity of the Dutch merchants, who joined forces in a company called the Vereenigde Oost-Indische Compagnie (Dutch East India Company). In its statute it was declared that this company enjoyed a certain number of privileges, such as the right to form an army, a police force and a judicial power, the right to establish a fort, the right to make agreements with the kingdoms of Nusantara, the right to issue currency, and the right to monopolize trade. The Dutch East India Company then went on to confirm its power through various agreements with local leaders. The form and content of the agreements varied considerably depending on the position and the state of the company. When the company was weak, for example, it simply sought protection from local leaders. Once the company had consolidated its position, it began to make demands, such as the right to negotiate monopolies. When the company felt it was in a position of strength, then the content of the agreement was to a greater extent imposed on the locals. Economic profit was the main reason for the violence orchestrated by the Dutch East India Company. These extreme measures can be seen in the logging of timber and the burning of clove crops in the Moluccas during the organized violence of the 'Hongi Expedition'.

The Dutch East India Company was able to arrange political 'contracts' with local leaders by taking advantage of internal conflicts among the 'Bumiputera' (local people), who eventually had to call on an outside force for help in order to defend the interests of their leaders. It could also be said that many agreements were originally determined by the residents themselves, or the local leaders. Thirdly, the Dutch East India Company built and maintained its power and its ability to cope with enemy attacks by attacking and subjugating local leaders. Attacks were organized that were intended to break down barriers to economic and political interests. Moluccas, Gowa, Banten, Banjar, and a number of other kingdoms were attacked by the Dutch East India Company. On the other hand, the Dutch East India Company was also attacked by various political forces from different kingdoms of the Indonesian archipelago. All the Dutch East India Company's centres of power, Ternate, Tidore, Bacan, Ambon, Gowa (Makassar), Banjar, Palembang, and Jambi in Aceh were attacked by local residents and leaders. Even in Batavia, the central government of the Dutch East India Company was attacked several times by Sultan Agung, the king of Mataram.

The Dutch East India Company functioned as a state within a state. The Dutch officer of the highest rank in Batavia was called the Governor General. Dutch local officials were appointed under various titles, such as Opperkoopman, Koopman, Kommissaris, Kommandeur, Opperhoofd, Gezaghebber and Commandant (Chief Merchant, Merchant, Commissioner, Commander, Chief, Officer and Commander). The operations of the Dutch East India Company were also supported by employees acting as traders.

In the first 20 years of its existence, shares in the Dutch East India Company went up by 300% and by more than 1000% a century later. During the existence of the Dutch East India Company, the average annual dividend distributed by the company was 18%. However, for a huge company with little or no shareholder control, whose shareholders were particularly satisfied with the amount of dividends they received, the VOC experienced serious financial problems in the last few decades of its existence. VOC officials diverted funds from their company through massive corruption. This corruption was one of the causes of the collapse of the Dutch East India Company. The VOC, which was the Dutch-language acronym for the company that had existed for about 200 years, was eventually used mockingly and was converted to stand for Vergaan Onder Corruptie (Perish Under Corruption). The Dutch East India Company or VOC was brought to an end in Nusantara in 1799.

∨ 90. Cannon (pierrier*)*

19th century
Indonesia
Bronze, 23 x 202 x 20 x cm
MNI 62

Cannon was first brought to Indonesia by Europeans in the 16th century. The inhabitants of the archipelago quickly understood the importance of this type of armament and either bought or took it from the enemy. As from the 19th century, they began to cast it themselves. The pierrier has a hole in its base to speed up firing. Cannon decoration had significance. This one resembles a dragon with an open mouth. Not only does this tell us that the weapon was attributed with supernatural powers, it also informs us that it was probably cast in a region of Indonesia influenced by Chinese culture, where the dragon was omnipresent.

> 91. *Padrao*

16th century
Jakarta
Stone, 172 x 34 x 44 cm / weight: 588 kg
MNI 26/18423

A *padrao* is a vertical stone used to mark Portuguese territories. This one commemorates the treaty signed between the kingdoms of Sunda and Portugal on 21 August 1522. Its upper section contains an armillary sphere crowned by a clover. In order from top to bottom beneath the sphere are the cross of the Order of Christ, the inscription OSPOR (O Senhor de PORtugal), the inscription ESFERRa/Mo (Espera do Mundo) – together this pair of inscriptions signify 'the king of Portugal [is] the hope of the world' – and lastly another cross. The treaty of 1522 was made in two copies; it stipulated the conditions under which Portugal, represented by Jorge d'Alberquerque, the governor of Maluku, would come to the aid of the king of Sunda, who felt his kingdom was threatened by the sultan of Demak. Drawn up by Captain Henrique Leme on one hand, and the crown prince Prabu Surawisesa Jayaperkasa on the other, the treaty of friendship benefited both parties. The Portuguese gained the monopoly on trade and authorisation to build a fortress to fight Sunda's Muslim enemies, for which Sunda would pay the king of Portugal an annual tribute of 1000 bags of pepper. It was agreed that the fortress would be constructed in the port of Sunda Kelapa on the right bank of the River Ciliwung. To give material form to the agreement, a *padrao* was erected. The fortress never saw the light of day: as all Portugal's available military forces were occupied in Goa, Fatahillah, the crown prince of the sultanate of Demak, took advantage of their absence to conquer the port, which he renamed 'Jakarta'. The fact that the Indonesian national holiday is on 22 June is because the port was conquered on 22 June 1527.

111

ᵛ 92. Map of Maluku

1729
Unknown origin
Paper, 45 x 34 cm
MNI 292

During the 15th century, when navigation made it possible for ships to sail around the world, the science of cartography advanced greatly and increased in precision. It was at this time that spices (cloves, nutmeg and pepper) were literally worth their weight in gold. One of the principal destinations for navigators was Maluku on account of its spices. Arab, Portuguese and Dutch merchants strove to win the monopoly on this trade. On this map, drawn up in 1729 in Leiden by Pieter van der Aa, a cartographer and printer, the form of the islands is not yet exact.

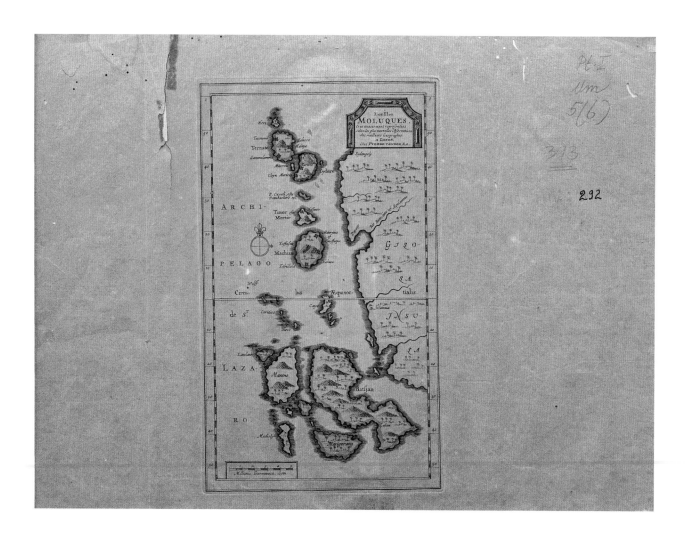

> 93. Dish

1521–1567 (during the reign of the
Jiajing emperor, Ming dynasty, China)
Seram Island, Maluku
Porcelain, 6 x 30.5 cm
MNI 2377

The bottom of this dish has a blue
underglaze decoration of three *hong*
or *fenghuang*, two names for the same
bird in Chinese mythology. During the
Han dynasty 2200 years ago, the male
feng and female *huang* were shown
face to face. From the time of the Yuan
dynasty, these confronted images were
transformed into a female *fenghuang*,
the symbol of the empress, and the
male dragon, the symbol of the emperor. The motif spread throughout
the archipelago, as is attested by the
batik of Lokcan (see cat. 137) and the
lacquered dish from Palembang (see
cat. 166) (de Flines 1969, Li Zhiyan 1989,
Ridho 1979, Vainker 1977).

< 94. Bottle

16th century (Ming dynasty, China)
Tobelo, Halmahera, Maluku
Porcelain, 25.2 x 12 cm
NMI 1706

The four globes on the body of the
bottle, which symbolise Portugal during the reign of Manuel I (1469–1521),
are inscribed at their centre with the
word *ESPERO* ('I hope'). They make
this object extremely rare. Artefacts
like this were only manufactured
in Fujian in China. Even if the bottle
is not exactly vertical, it reflects the
precision of the order made by Portuguese merchants who had settled on
the 'spice island' responsible for their
wealth (de Flines 1969; Ridho 1979;
Vainker 1997).

^ 95. Dish

18th century (Edo period, Japan)
Lampung, Sumatra
Porcelain, 10 x 54 cm
NMI 3871

The motifs on this dish – the eagle, rock and cherry tree on the flat bottom, and the small bird, rock and fan on the inner walls – are of Japanese origin. They are executed in blue underglaze enhanced with red, gold, black and white overglaze. In Japan, a porcelain industry developed in the city of Arita during the 17th century following the discovery there of kaolin. Beginning in 1659, its production was exported to Europe and the Arabian Peninsula from the port of Imari by the Dutch East Indies Company. This development was the outcome of the decline in the Chinese porcelain industry subsequent to dynastic disputes that brought the Qing to power in 1644; also to the growing demand for luxury products not only in the archipelago but in Europe, where increasing prosperity went hand-in-hand with changes in taste (Adhyatman 1982, de Flines 1969, Ridho 1979, Vainker 1977).

114

96. Bottle

18th–19th century (Fustat-Misr, Cairo)
Kudus, Central Java
Faïence, 28 x 15.5 cm
MNI 2759

Decorated with engraved and applied motifs, this greyish-brown unenamelled bottle was used to carry holy water after a pilgrimage to Mecca.

97. Jarlet

16th century
Gowa, South Sulawesi
Porcelain, 12 x 11.5 cm
MNI E 550/12532

Small and round, this cream-coloured object was made in thick porcelain in Vietnam. The slightly crackled surface is divided with blue underglaze into vertical rectangles and covered with repoussé gold-leaf *kawung* motifs

(a stylised four-petal flower). Gold and silver leaf began to be used on porcelain in the 16th century to hide crackling and/or enhance the value of the object. The latter was certainly the purpose for this small jar as, according to records, it was part of the regalia belonging to the sultanate of Gowa (Adhyatman 1982, de Flines 1969, Ridho 199, Sundari 2014, Vainker 1977).

∧ 98. Dish

19th–20th century
West Java
Faïence, 26 cm
NMI 9346

This dish is embellished with Arabic inscriptions in blue verging on mauve. Earlier Chinese wares of this nature (16th–17th centuries) were difficult to understand as the craftsmen had not necessarily mastered the Arabic language. The printing technique used on this dish made in the Nether-

lands improved this situation. The inscription in the middle reads 'Allah does not forgive association with Him [Quran 4:48]. Muhammad is his messenger. Then, you will be helped'. On the rim there are two instances of the inscription 'There is no deity except Him' [35:3], which is intersected by the names of the Rashidun Caliphs: Abu Bakr, Uthman, Umar and Ali (Adhyatman 1982, de Flines 1969, Jorg, 1983, Ridho 1979, Vainker 1977).

116

> 99. Dish

17th century (Ming dynasty, China)
Wonosobo, Central Java
Porcelain, 6.5 x 33 cm
MNI 3767

This attractive but rather poorly finished dish (it has small holes formed by the water that evaporated during firing) comes from the mass-produced pieces made in Zhangzhou. The very simple ornamentation means 'Praise be to Allah, Lord of the universe' (Quran, 1:2). According to Orsoy de Flines, this dish was still being used in religious ceremonies in 1936.

> 100. Dish

17th century (Ming dynasty, China)
Martapura, South Kalimantan
Porcelain, 11.5 x 48 cm
MNI 2129

The lack of finesse in this dish is due both to the materials used and a lack of monitoring during the firing. It is one of the products exported in vast quantities to South-East Asia from Zhangzhou via the port of Shantou (Swatow). The dish is decorated with images of European ships and fish.

> 101. Dish

17th century (Ming dynasty, China)
Ternate Island, Maluku
Porcelain, 8 x 35.5 cm
MNI 3602

This blue underglaze dish has a motif associated with the Taoist 'islands of the immortals' or represents international sea traffic in a port like Guandong (China). As the VOC (Verenigde Oost-Indische Compagnie, or Dutch East India Company) was unable to buy directly in China, it purchased these wares in the ports packed with Chinese junks of Jakarta and Ternate, the 'spice island'.

‹ 102. VOC bottle

18–19th century (Netherlands)
Banten, West Java
Stoneware, 25 x 18 cm
MNI 2780

Made from thick, pale grey stoneware and covered with pale brown and pale blue enamel, this bottle is decorated with applied motifs of lotuses and a bearded face, for reason of which it has been named the 'bearded man bottle' ('*baardmanfles*' in Dutch). It was used to hold alcohol.

‹‹ 103. Treasure from the *Tek Sing* shipwreck

Most of the 350,000 objects salvaged in 1999–2000 from a junk discovered in the Gelasa Strait were Chinese Qing-dynasty wares (1644–1912). It is called the *Tek Sing* cargo, after the name of the ship (which means 'true star'), a traditional Chinese junk measuring 50 metres long by 10 metres wide, which was able to carry 400 crew-members and 1600

passengers. The junk left the city of Amoy (today Xiamen) in Fujian province on 14 January 1822 on the way to Batavia (today Jakarta) in Java. It never reached its destination: it sank in the Gelasa Strait on 5 February on account of bad weather.

∧ 104. Glass

Probably 18th century
Glass, 23 x 16 cm
MNI 8471/127

Decorated with the VOC monogram (Dutch East India Company), this glass was probably a gift or used as ornamentation.

∧ 105. Wine glass

17th century (Netherlands)
Crystal, 22 x 9 cm
MNI 7462/124

This wine glass is decorated with a Dutch sailing ship and bears an inscription meaning 'Dutch East India Company. Prosperity'.

120

ᵛ106. Oil lamp

Unknown origin
Bronze, 18 x 24 x 15 cm
MNI 1103

Either hung or standing, this lamp
was used in religious ceremonies.

121

THE MODERN PERIOD

FROM THE NINETEENTH CENTURY TO THE PRESENT DAY

WESTERN COLONIALISM

Shortly after the collapse of the Dutch East India Company, the territories that the company had assembled were taken over by the Batavian Republic. The Indonesian archipelago became a colony. In order to manage this colony, the Dutch government in Europe appointed a governor general. Albeit reluctantly, the Dutch colonial government sought to renew and modernize the colonized country in an attempt to secure reasonable benefits for themselves. From 1811 to 1816, Indonesia was occupied by the British. As the highest-ranking British leader in Indonesia, Lieutenant Governor General Raffles also tried to bring about a number of reforms and modernizations. The trend for policies aimed at modernization lasted about 30 years until a system of exploitation began to be applied in 1830, using one of the policies that had been used by the Dutch East India Company, namely that of the Cultuur Stelsel or 'Cultivation System'. (In contemporary Indonesian history, it is called Tanam Paksa (enforcement planting).

However, after the first 25 years, the Dutch colonial government began to adopt monopolistic methods. In 1825, for example, the Dutch colonial government allowed the exploitation of the Netherlands Trading Society (Nederlandsche Handelsmaatschappij, or Netherlands Trading Society) to engage in exploitation in Indonesia. The company was set up under the decree of the Dutch King to 'increase trade, shipping, the shipbuilding industry, fishing and agriculture [...] to build roads, bridges, ports and warehouses [...] and to bring great benefits to the state'. In a short period of time, this company had established its agency in Batavia, Semarang, Surabaya, Padang, Makasar and various important cities in the Dutch East Indies.

The Netherlands Trading Society can be seen as an extension of the Dutch government in Europe. The government granted the company the rights to sell and purchase various trading products and the right to transport and distribute a variety of government products. In addition, the government provided protection against threats from rivals. In other words, the Netherlands Trading Society emerged as a monopolistic society, so the Dutch acronym for the name of the company, the NHM, was often referred to as 'Niemand Handelt Meer' ('Nobody Trades More').

The monopolistic nature of the Dutch government of the East Indies became more obvious when the Cultivation System was introduced in 1830. The cropping system is an economic policy that obliges all farmers to plant a number of cash crops on their lands, to take care of the crop and to sell the products at a price determined by the government. Some of the plant types that the farmers were forced to grow were coffee, sugar cane, tea, tobacco, cinnamon and cotton.

The introduction of the Cultivation System triggered the development of transport infrastructure and facilities as well as other economic institutions, and the government built a lot of roads, bridges and ports and introduced the use of cikar (two-wheeled carts drawn by oxen), pedati (horse-drawn carts), and boats to transport the products. Many entrepreneurs and enterprises, generally European, from the Far East or from the Bumiputera (local populations) made it possible for the Cultivation System to operate, a policy that allowed the emergence of new groups of entrepreneurs protected by the government.

In the middle of the nineteenth century, a political change took place in the Netherlands.

Liberals began to dominate parliament, which had a major impact on economic policy in the colonies. A private sector started to emerge within the economy, which had until then been controlled by the state. Many companies therefore began to invest in Indonesia, particularly in the period after the 1870s. Various plantations, and then mines, were opened in Indonesia. In order to facilitate the mobility of people and goods, more and more roads, railways and ports were built and improved, and new modes of transport were introduced, such as trains and steamboats.

The second half of the nineteenth century was an important era for the Indonesian transport system. Trains were being used on land, while steam-powered transportation was gradually being introduced at sea. These modes of transport also modified the tonnage of goods that it was possible to transport. Economic and political changes took place in correlation with the political expansion of the Dutch colonial government beyond Java. These attempts at conquest were made through military expeditions throughout the whole of Indonesia. Through the superiority of their weapons and the use of divide and rule policies among the various groups in Indonesian society, the Dutch succeeded at the beginning of the twentieth century in uniting Indonesia into the 'unitary state' of the Dutch East Indies.

Military expeditions at sea were carried out in various regions. During the first few decades, the Dutch government used its own warships, sometimes with the support of ships owned by the Netherlands Trading Society. The government also used vessels belonging to state-subsidized shipping companies, such as Cores de Vries, NSM (the Netherlands Steamship Company), and KPM (the Royal Packet Navigation Co.

Due to the scattered nature of the islands in the region, the Dutch East Indies Government created a Department for the Navy and War. Since the mid-nineteenth century, the average number of warships per port has been three times the average for small and medium-sized ports, and five to seven times the average for major strategic ports.

As the demand for ships grew, the government allowed the private sector to establish shipping companies. The government also granted privileges in the form of monopoly shipping rights in the archipelago and protection against rivals, as

well as generous subsidies. From the middle of the nineteenth century, the Dutch Government of the East Indies greatly increased the number and the quality of the ports in Indonesia in order to support shipping activity. The Government also took a number of steps to improve transport by sea, such as the construction of lighthouses, or the preparation of maps and navigation manuals for crewmembers.

In line with the policies described above, ships belonging to the bumiputera (local populations) were limited in terms of movement and size. Traditional shipbuilding industries were also strictly controlled. This control was carried out by limiting timber supplies and the levels of production. As a result of this policy, the number of local vessels decreased considerably. The services provided were also reduced to local or inter-island transport services, and more and more people engaged in activities related to the sea were forced to change profession. The presence of subsidized shipping companies made sense for the Pax Nederlandica of the Dutch colony. The boats of the shipping company were obliged to serve all of the Indonesian islands, which meant that the presence of these companies was also destined to unite the archipelago.

The government's hegemony in Indonesian waters was not only supported by the presence of warships from the Dutch Ministry of Defence and its 'national' ships, but also incorporated into the rule of law through the issue of the Territorial Sea and Maritime Circles Ordinance. This legislative framework, promulgated in 1939, regulated the territorial waters and maritime environment of Indonesia. One of the highlights of this order is the provision that the territory of Indonesia only extended three nautical miles from the lowest point of an island. This provision led to the existence of a free maritime territory in the Indonesian islands. In order to maintain the safety of the seas, the Dutch Government of the East Indies paid special attention to the Ministry of Defence and to the existence of its national shipping company.

Political expansion and economic exploitation by the Dutch government of the East Indies, particularly in the shipbuilding and shipping industries led to resistance from the bumiputera (local populations). The most conspicuous form of resistance was the 'boat cruise', which was intended to resist the bans put in place on the local sailors by the government. 'Boat cruising' was mainly carried out through the use of small craft carrying a variety of goods to be exchanged in cities outside territories dominated by the Netherlands – such as Penang Island and Singapore. The other reaction was to lead pirate-style attacks on the commercial vessels protected by the government. The tradition of attacking colonialist ships was in fact a continuation of an earlier tradition developed and practiced by the bumiputera (local people) and dating back to the era of the Dutch East India Company, a tradition that only recently came to an end.

THE PACIFIC WAR

Not many history books have mentioned the Japanese presence in Indonesia before the start of the Pacific War. The earliest historical sources suggest that the Japanese presence in Indonesia begins in the seventeenth century. In 1623, there were reports of up to 30 'Japansche burgerij' (Japanese bourgeoisie) engaged in maritime trade in Batavia. In addition, a number of Japanese were also present on the island of Banda (unfortunately, it is not known exactly how many) at the time of the Dutch East India Company. With the policy of Sakoku (Isolation), the Japanese disappeared from Indonesia in the period from the eighteenth century until the final quarter of the nineteenth century.

The Japanese began to reappear in Indonesia in the last few decades of the nineteenth century. The first demographic data on the Japanese, which appeared in 1875, revealed that there were up to 463 of them in Indonesia. 376 of them were women and they generally worked as prostitutes. At the beginning of the twentieth century, the number of Japanese increased. Their professions were generally associated with the world of trading, spices and fishing.

In addition to the growing number of Japanese settlers in Indonesia, a number of Japanese shipping companies also established operations in Indonesia. Despite fluctuations in the services provided, dozens of Japanese shipping companies settled and operated in Indonesia in the period leading up to the 1930s, including: Nanyo Yusen Kaisha, Osaka Shosen Kaisha, Nippon Yusen Kaisha, Taiyo Kaiun Kabushiki Kaisha, Kawasaki Kisen Kaisha, Mitsui Busan Kabushiki Kaisha, Katsuda Kisen Kaisha, Kokusai Kisen Kabushiki

Kaisha, Toyo Kisen Kabushiki Kaisha, Yamashita Kisen Kabushiki Kaisha, Tatsuuma Kisen Kabushiki Kaisha, Nippon Kisen Kabushiki Kaisha, and Kaishi Kaishi Kaisha. These Japanese shipping companies had contacts with the principal ports of Indonesia, such as Belawan (Medan), Palembang, Teluk Bayur (Padang), Tanjung Priok (Jakarta), Semarang, Surabaya, Makassar, Manado, Samarinda, Tarakan, and a number of other small ports.

As well as commercial vessels, Japanese fishing vessels were also present in Indonesian waters, although their activities were concentrated in the waters of the Moluccas and Riau. Later, when Japan entered Indonesia, these two areas became important domains for Japanese military operations. And when Japan had power over Indonesia, it was understood that most traders or fishermen were also part of the troops of Dai Nippon, the Greater Japanese Empire.

MARITIME ACTIVITIES AND THE GOVERNMENT OF THE REPUBLIC OF INDONESIA

On 10 September 1945, the former Navy cadets and former pupils of the Jakarta Maritime College (Sekolah Pelayaran Tinggi Jakarta) were assimilated into the city's central branch of the Maritime Security Service (BKR Laut), which was led by Mas Pardi and based in Jakarta. This led to three bases being set up: the West Java base in Jakarta, the Central Java base in Semarang, and the East Java base in Surabaya. On 5 October, the BKR Laut was converted into the Navy for the Protection of the Population (TKR Laut). Because of a fear of the Netherlands Indies Civil Administration (NICA), the headquarters of the TKR Laut were transferred from Jakarta to Yogyakarta and consisted of two main divisions: the West Java division at Cirebon and the Java Central division at Pursiegeworejo. Meanwhile, the East Java base was placed under the charge of the Yogyakarta headquarters. In Sumatra, the TKR Laut was headed by Suhardjo and it was based in Palembang, before being moved in succession to Siantar, Prapat, and Bukit Tinggi. The TKR Laut of Sumatra was officially declared part of the Navy of the Republic of Indonesia (ALRI) on 19 July 1946, with a number of bases including Tanjung Balai (East Sumatra), Palembang, Tanjung Karang (Lampung), and Padang (West Sumatra).

Creating Indonesian naval bases outside Java and Sumatra was very difficult because these territories had not been recognized by the Dutch in the Linggajati Agreement of 1946. As a result, a number of military expeditions were carried out. The expedition to Borneo was conducted from the ports of Tegal, Surabaya, and Pekalongan. After a long drawn-out process, the ALRI Division IV of South Kalimantan was created on 17 May 1949 with command over all the Navy bases in Kalimantan and the division was placed under the leadership of Hasan Basry. The military expedition and efforts to establish naval bases in Sulawesi were much more problematic. This was not only because the Dutch still exercised a lot of power, but also because of the terror spread by the forces of Raymond Westerling, which had led Indonesian fighters to move to Java. The State of East Indonesia was then established, with Makassar as its capital.

Efforts to assert sovereignty continued. In the Cabinet of Djuanda, two maritime decisions were taken that are worthy of note. First was the nationalization of the Dutch shipping company (KPM) on 3 December 1957. The national shipping company (PELNI) continued with the task of providing inter-island transport. The second was the extension of the Territorial Sea and Maritime Circles Ordinance of 1939 from a 3-mile boundary to a 12-mile boundary (Djuanda Declaration, 13 December 1957). The latter policy brought about the expansion of Indonesian territory, the end of the enclaves, and the 'freedom' of the sea routes – a paradigm shift in terms of the sea, which was no longer seen as something that separates, but rather as something that connects. In the 1960s, the Djuanda Declaration was incorporated into law and onto the platform of the Wawasan Nusantara (Archipelago Concept), which was accepted by the United Nations in 1982.

The bravery of the Indonesians in navigating the oceans was highlighted in 1986 at the Phinisi Nusantara exhibition at the Marine Plaza in Vancouver. The boat was built at Tanah Beru Bulukumba, under the direction of the panrita lopi (ship surveyor) Haji Damang and the work was completed at the IKI Ujung Pandang shipyard. The vessel left the port of Muara Baru Jakarta and sailed through the Java Sea, the Flores Sea, the Banda Sea, the Maluku Sea, the Halmahera Sea,

and across the Pacific Ocean to Victoria Harbour. Skipper Gita Ardjakusuma led the 11 crewmembers for the duration of the 69-day cruise (9 July to 14 September) and for a distance of 11,600 miles. After the exhibition, the ship Phinisi Nusantara was handed over to California Universities Sailing Inc. (CUSI). A model of the Phinisi Antarbangsa that was also exhibited at the event was similarly donated to the Maritime Museum of Vancouver. Both the ships became seagoing Indonesian ambassadors to Canada.

30 years after the expedition of the Phinisi Nusantara, the Indonesian government sent another expedition to Japan in 2016. Some 20 metres long and 4.5 metres wide, *The Spirit of Majapahit* was built in Selopeng Beach, Sumenep (East Java) in 2009. Skipper Muhammad Amin and a crew of 10 left for Japan from Pier 21 Marina Ancol in Jakarta, via the Jakarta-Pontianak-Brunei-Manila-Taiwan-Japan sea route.

This ship made two attempts at the adventure, but sadly without success. On the first crossing in 2010 the ship had to turn back after being struck by a typhoon in the waters of the Philippines. The second cruise was planned for 2011, but had to be postponed due to the tsunami in Japan. The strong desire to carry out the expedition was based on historical values. In the thirteenth century, the ships of Majapahit reached Japan on several occasions, which is indicated by the discovery of a kris from Majapahit in Okinawa. In the opposite direction, fragments of Japanese Imori ceramics were found on the site of Trowulan in the east of Java, which was the location for the Kingdom of Majapahit. The ship used for this expedition was therefore designed according to the ships depicted on the reliefs at the temple of Borobudur and on models of Majapahit ships, which is why the expedition was known as the Ekspedisi Kapal Majapahit (Majapahit Naval Expedition).

LIFE IN THE MARITIME COMMUNITIES

As a maritime country, Indonesia was culturally enlivened by several maritime tribes. Historical and cultural studies have unearthed evidence for the existence of six main tribes, namely the Bajo, Bugis, Makassar, Buton, Mandar, and the Madura. The Bajo themselves have several tribes: the Tribe of the Sea, the People of the Sea, the Sama, the Bagai, and the Boat People. In the past they lived nomadically on boats. Most Bajo tribes now stay on land near the sea, or on small islands. Their principal activity is to provide the people living on the mainland with products from the sea.

The Bugis and the Makassar settled on the south coast of Sulawesi. The centre of the shipbuilding industry is to be found in Bulukumba. Different types and sizes of ships and boats, including padewakang, palari, pinisi, and lepa-lepa are built there. Besides fishing, the tribes of the Makassar produce salt in Jeneponto. The Mandar tribe lives on the west coast of Sulawesi. They produce a variety of boats such as pakur, sandeq, padewakang, baqgo, lambo, and lete, in Pambauang, Pambusuang, Majene, and Campalagian. The Buton tribe lives in the southeast of Sulawesi. Their main trading boat is the bangka/wangka, which is also called lambo or boti. Other types of boats include the jarangka (double rocker) and the kole-kole, which are used to catch fish and carry short-distance cruises. The most famous sailors came from the Wakatobi Islands, especially from Binongko Island. The Madurais live mainly in the island of Madura and the surrounding islands. They manufacture and use many types of boats, including janggolan, lis-alis, golekan, and leti-leti. Besides fishing and trade, this tribe also take part in the mining of salt.

There are other tribes in Indonesia besides these. The Biak tribe is located in Cendrawasih Bay in Papua. The inhabitants of the Moluccas have their own typical boat, namely the kora-kora (belang). The inhabitants of Tanimbar finish the tip of the stern and the prow on their boats with decorative patterns. As well as engaging in fishing, the inhabitants of the Moluccas cultivate pearls, particularly the inhabitants of the islands of Banda and Kei.

In general, maritime tribes use different types of fishing gear, such as fishing rods, nets, rumpon (fish aggregating devices, or FADs), bagan (lifting nets) and sero (guiding barriers). The fishermen of Lamalera in the east of Nusa Tenggara have a unique tradition of whaling, which is not just a question of catching what they need to live, but also of maintaining their cultural values. 10 metres long and with a convex hull of 1.5 to 2 metres, their boat, which is called the peledang, is a tribal symbol. Fishing activities are carried out in the dry season (lefa season) from May to September, in the waters of the Flores Sea and the Sawu Sea. They only hunt for kotaklema whales (sperm whales). Although another type of whale, the klaru whale, is often seen, they are not fished because it is believed that the klaru whale has mythological links with the people of Lamalera. The klaru were used by the ancestors during their first visit to Lamalera. Hunting begins with a ritual led by the boat owners (Lamavujon tribe). The hunting is then led by a hunter (lamafa) and 10 to 14 team members (matros) of the crew. The whales that are caught are about 15 metres long and 2 metres high, and they are cut into 15 pieces that are then shared between the lamafa, the matros, the owners and the villagers. When the whales fail to appear during the lefa season, it is a sign that there are problems in the village that need to be resolved.

Almost all the activities of the maritime tribes are accompanied by ritual. For them, the sea is the place of life and the land is the place to take shelter. The sea is not an object, but a subject. The balanced relationship between man and the sea has to be maintained through Sedekah Laut (maritime offerings). The fishermen of Bone Bay call this the Maccera '*Tasi*' ritual, and it is organized at times of diminishing prey. The ritual led by the traditional chiefs (Puang Puawang) is therefore a form of respect for the Sustenance Giver, in gratitude for continuing to provide them with the means to live well.

As well as its cultural resources, Indonesia is rich in terms of sites for seaside tourism. There is the island of Sabang at the extreme western end of Sumatra and he island of Cubadak, a tourist resort on the west coast of Sumatra, is known as Paradiso Village. The region of the Sunda Strait has the Ujung Kulon National Park, and the Java Sea offers maritime tourism in the national parks of Kepulauan Seribu and Karimunjaya. Other maritime tourist sites are present in the national parks of Bali, Lombok, and Komodo in the Nusa Tenggara Islands. The national parks of Wakatobi and Bunaken are also located in Sulawesi. In the middle of the Banda Sea, there is a very beautiful group of islands, namely the Bair Islands, which are also known by the Moluccas name of 'Raja Ampat'. Two tourist attractions in the eastern end of Indonesia are the Raja Ampat National Park and Cenderawasih Bay in Papua.

THE PADEWAKANG: TRADITIONAL BOAT OF SULAWESI

SHORT HISTORY OF THE PADEWAKANG

The padewakang is a typical Sulawesi boat that was used for maritime trading by Bugis, Makassar, and Mandar seafarers from the seventeenth century until the beginning of the twentieth century (CE). This boat has an enlarged pajala-style body. It has a square stern, double wheels, a bow, and two jibs. Its stern has a three-legged mast with rectangular sails. The boat has a capacity of between 20 and 50 tons and the smaller version is called a palari. At the end of the nineteenth century (CE), the padewakang evolved into the pinisi, which is propelled both by sails and an engine.

The shipping network on which the padewakang was used travelled eastward to the Moluccas, and toward the western coast of Papua (Irian), where it was particularly used for the purposes of the spice trade during the seventeenth and eighteenth centuries (CE). The network extended southward to the northern coast of Australia, and this part of the network was used for the trading of sea cucumbers, the most sought after commodity for Chinese merchants in Makassar. In his journey note published in 1792, Thomas Forrest mentions that the padewakang boats often sailed to New Holland (Gulf of Carpentaria) to collect sea cucumbers (trepang) in order to sell them to the Chinese junks that came every year to Makassar. This particular type of sea cucumber was a premium commodity at the time.

Heading north, the network reached the Celebes Sea. A travel note from the eighteenth century mentioned that about 14 to 15 padewakang boats from Pasir (East Kalimantan) traded in the Sulu Islands every year. Some of those boats went to the Moluccas to buy spices that they then sold back in Sulu. From there, the boats headed to Batavia, Malacca, and Penang. Datus (regional leaders or elders) from Jolo bought gunpowder and weapons from Bugis and Mandar merchants. Meanwhile, the westward network reached as far as the Strait of Malacca and the Indian Ocean. Some local newspapers from the nineteenth century mentioned padewakang from Mandar carrying woven fabrics and dried fish to the west coast of Sumatra every year.

For centuries, the padewakang boat played an important role in the development of the Java Sea maritime network (the 'marine core' of Nusantara). The technology of the padewakang boat also initiated the development of the lete-lete, the trading-boat of Madura. The latter was also used by Mandar seafarers at the beginning of the twentieth century. The upper interior of the boat is divided into compartments for storing merchants' goods and as resting places for the boat crew. The skipper (juragan) would take the largest of these compartments. Boat owners received 5% of every merchant's profit as a lease. This method ended around the time of World War II.

THE BOAT-BUILDING PROCESS

The centre for padewakang boat-building in Sulawesi is in Bulukumba (South Sulawesi). Generally, the boat-building process is divided into three stages. First comes the search for wood in the forest by a group of around half a dozen, made up of the *punggawa* (patron) and *sawi* ('clients'). Vitec cofaccus, the type of wood sought for the keel and the hull, is always submerged in water and is called *bitti* in Buginese, or katondeng in Makassar. The other wood types required for the remaining parts of the boat are teak (*Tectona grandis*) and cembaga (*Petrocarpus indicus*). These woods are obtained from the forests in South Sulawesi, and are also ordered from Kalimantan and Southeast Sulawesi.

Second comes the building of the boat in the *bantilang* (shipyard). This stage is divided into six steps: (1) the making of the keel (*kalebiseang*); (2) the making of the hull; (3) the installation of *lammah* board; (4) the installation of the frame (stringers and planks); (5) the installation of *katabang board*; and (6) the installation of *ambeng* and *anjong*. The manufacture of the keel is led by the punggawa. At the beginning of the process, the punggawa is wrapped in white cloth by one of the owners. A chip taken from the first blow in the chopping of the wood is taken by the retainer, and then halved: one part being for the owner and the other for him. The owner puts the chip in a bottle filled with coconut oil. No longer covered in white cloth, the punggawa then continues the work until holes are formed at the base and the ends of the keel, which are then connected with the front and rear connector. The blood of a rooster and a hen are then sprinkled over the two connections, symbolizing intercourse between husband and wife.

Third comes the preparation for the launch of the boat into the sea. This begins on land with the reading of *barzanji* and songko-bala to ward off disaster, the shaping of the 'navel' (*pamossi*) of the boat, and the offering of food at night to the supernatural inhabitants of the hull (*apakanre-ballapati*). The shaping of the navel is led by the punggawa. He is always wrapped in white cloth when drilling the keel. Two *sawi* wait under the hull and collect the debris from the drilling. The debris is then given to the boat owner and put into the bottle of coconut oil, together with the preserved chip from the first chopping of the wood for the hull (*kalabiseang*). The panrita (*teacher*) then proceeds to make an offering of food to the supernatural spirits. The boat owner sits to the right of the panrita and the skipper to his left. Next, a hatched chick is cut and mixed with pieces of banana flower, which is then wrapped with banana leaves and put in certain parts of the boat. The wife of the boat owner brings out complementary food, which is accompanied by the recital of a prayer by the panrita and is later placed on the front and rear keel connectors. The next morning, the boat is launched onto the sea with the priest leading proceedings, assisted by members of the local community. One of the community members is in charge of hitting the gong standing near the boat to encourage those charged with getting the boat into the water. When the boat begins to move, the wife of the boat owner and some other women strew rice onto the boat. This ritual ends with the taking of a communal meal together on the beach. All of those stages are considered to be equivalent to the process of human birth, so the boat is treated like a 'human' who needs to be properly looked after for the sake of his or her survival.

ABD. RAHMAN HAMID

129

< 107. Prow ornament

Banjarmasin, South Kalimantan
Wood, 115 x 66 x 45 cm
MNI 2576b

Coloured prow ornaments like this
one adorned the boat the king used
during his journeys of inspection.
In this one, two heads are combined:
those of a dragon and an elephant.
Associated with both earth and water,
the dragon is the companion of a goddess who symbolises vitality, protection in daily life, fertility and rebirth
after death. The elephant represents
the Hindu god Ganesh, who is reputed
to remove obstacles.

> 108. Paddle

Papua
Wood, 127 x 11.5 cm
MNI 18021c

Among the Amat, wood carvings are
a means of communication between
the living and the dead, between the
worlds of the spirits and humans.
From early childhood, boys learn to
work wood, from tree-felling to the
making of tools and the representation
of the ancestors. On the north-east
coast of Papua, the motifs most often
carved are of fish, birds and ancestor
figures, all of which are linked with
protection and the community's
dependence on water as the source
of life and the medium of transport.
Dugouts used by women for fishing
on Lake Sentani are powered and
guided by paddles magnificently
carved with motifs of fish created
by combinations of S-shaped spirals.

>> 109. The Nogowarno horse

Cirebon, West Java
Wood, 160 x 130 cm
MNI 8488

A woman stands on a winged creature
with the body of a horse and the
crowned head of a dragon. The sculpture stands on a plinth decorated with
traditional puppets (*wayang*). The
horse represents *Buraq*, the mount
ridden by the Prophet. The female
figure is Nyai Roro Kidul, a goddess

who plays an important role in Javanese society. Her name means 'goddess
of the South Sea' (=of the Indian
Ocean). She is associated with water
and the world of the spirits, therefore
with the forces of life and death and
fertility. She is worshipped for her
generosity but also feared for the
devastation she is capable of wreaking.

>> 110. The sun god Upu-lera

Leti Islands, Babar Islands
(South-West Maluku)
Wood, 212.4 x 67 x 11 cm
MNI 14304

This sculpture is a representation of
the sun god. The inhabitants of the
islands of Tanimbar, Babar and Leti
believe in the existence of a god who
can take different avatars, namely
Upu-lera (the sun), Lor Wol (Sun-
Moon twin god), and Wuli and Lojtien
(the gods of animals, who are often
evoked in local songs), all the celestial
gods (Romuty 1967: 18).
The presence of the sun god in the
village was also manifested in the
lamp made from coconut tree leaves
and hung in front of the house or in a
sacred tree. The god would descend
from the tree once a year down the
wooden ladder placed against it at the
start of the rainy season, so that he
might spread fertility.

ᵛ111. *Omo Hada* from the south of Nias Island

South Nias, North Sumatra
Wood, bamboo, 102 x 118 x 58 cm
MNI 21260

The traditional houses in the southern part of Nias Island are called *omo hada*. The inhabitants of this part of the island believe that their ancestors are superior beings who migrated to the east coast of the island and then settled at the mouth of the River Gomo. These ancestors quickly learned the dangers of their new environment: earthquakes and tidal waves. To protect themselves, they reminded themselves of the dangers they had overcome in their boats and took this inspiration to built houses that are sufficiently solid yet also flexible enough to resist earthquakes.

> 112. Spoon

Tanimbar Islands, Maluku
Mother-of-pearl, 14.5 x 4.5 cm
MNI 6754

The inhabitants of the Tanimbar Islands are instinctively attached to the sea. Shells play an important role in their community life: they not only provide nourishment but can also be transformed into a musical instrument or object of daily use, like this spoon made from a section of nautilus shell that has been cut and polished. Some spoons of this kind are engraved with sacred, profane or simply decorative motifs. This one illustrates the arrival of Europeans at Tanimbar on large steam-powered ships.

134

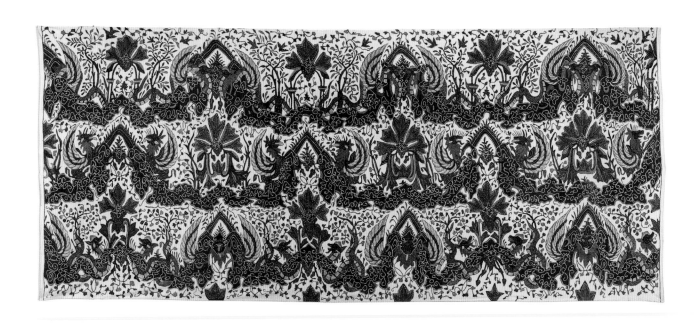

<< 113. *Tampan* cloth

Lampung, South Sumatra
Cotton, beads, pearls, rattan, 104.5 x 60 cm
MNI 577

Cloths made on Lampung are distinguished by their central motif: a boat, which symbolises the passage of life. They are of three types.
1. The *tampan* (or *nampan*). Generally less than a metre long; used in rites of passage to cover marriage presents (*seseharan*) or, in the case of the *tampan maju* (made of special beads), to cover the bride's seat.
2. The *tatibin*. Usually about a metre long; used as a wall decoration and as a *seseharan*.
3. The *palepai*. Generally more than three metres long; hung on the wall during rites of passage. Only village chiefs had the right to a *palepai*. This one was bequeathed to the eldest son but could also be lent to other family members.

<< 114. Batik

Cirebon, West Java
Cotton, 250 x 104 cm
MNI 30510

The motifs used in this cloth – *mega mendung* (see cat. 115), *garuda* (mythological animal, half-horse and half-eagle), lion and tree – are all of Chinese origin and typical of Cirebon royal batiks. Their common themes are life, fertility and the battle between good and evil. The *barong* (an animal derived from a lion that elsewhere in Indonesia is called a *singo*) chases away evil spirits and battles against the forces of the underworld. As the *barong* is linked to the world of water, it is also associated with the fertility expressed by the *mega mendung* motif, as it is clouds that allow rain to fall to fertilise the earth and bring life to nature. The dark blue represents the rain-bearing clouds and the pale blue life purified by the rain. The tree too is a symbol of life. As for the *garuda*, it has been influenced by motifs originating in Solo and Jogja.
The theme of life is understood as the battle between good and evil by all Indonesian cultures and religious traditions. In Cirebon, a city with multiple cultural influences, it is not only the batiks that reflect this universal theme. It is also seen in the decoration of the sultan's palace where the *garuda* is a reference to the *Buraq* (the Prophet's mount during his night-time flight to Mecca [see cat. 115]) and thus to Islam, the elephant to the Hinduism of East Asia and India, the dragon to Chinese Buddhism, and the lion to Protestantism and European culture.

<< 115. *Mega mendung* batik

Cirebon, West Java
Cotton, 272 x 105 cm

The *mega mendung* is one of the batik motifs that are the prerogative of the royal family of Cirebon. Seen on ceramics and cloths, the representation of clouds was brought to Cirebon by the Chinese. Following the marriage between Sunan Gunung Jati, one of the nine Muslim saints revered in Indonesia, and Ong Tien, daughter of a Ming emperor, many Chinese formed relations with the region as from the 15th century. Referring to the world above in Taoism and many other cultures around the world, and integrated by the Sufis into their artworks, the motif was also taken up by the artists of Cirebon. It is created using either five or seven colours. The number 5 symbolises the pillars of Islam, and 7 the number of heavens that the Prophet passed through during his night journey (Surah 17 of the Quran). If the *mega mendung* has more colours, it is all the more precious on account of the increased technical difficulty.

> 116. *Dewaruci*

Yogyakarta
Leather, horn, 21.5 x 47 cm
MNI 30710

Bima is an honest, brave and faithful warrior, the hero of *Dewaruci*, a book in which he is sent in search of holy water. In his quest at the bottom of the ocean, he becomes gravely wounded. On regaining consciousness, he finds he has been washed up on the shore of an island. It is there that Dewaruci tells him that holy water is found deep inside oneself, in the place where the battle between good and evil takes place. Armed with this knowledge, Bima is stronger than ever. An outstanding warrior, he devotes himself to the defence of his country.
Given this background, it is unsurprising that the Indonesian Navy's training ship, launched in 1953, is named 'Dewaruci'.

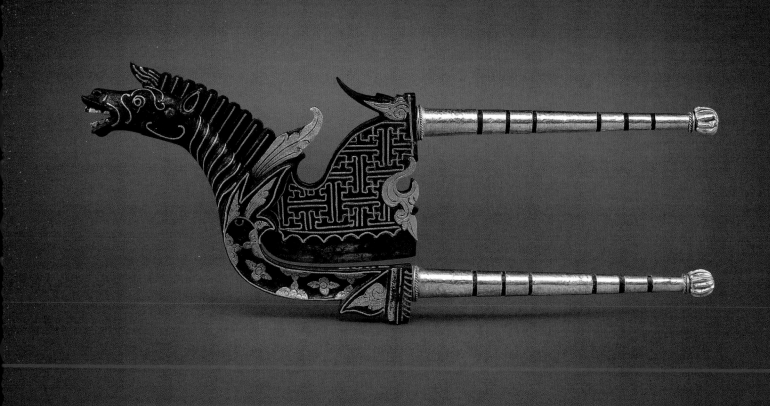

< 117. Betel-nut scissors (*kacip*)

Lombok, West Nusa Tenggara
Gold-plated iron, 8.5 x 20 x 17 cm
MNI E. 1020/8008

As a result of its maritime trading, in the 17th century Lombok became one of the richest kingdoms in southern Indonesia. The gold and silver objects in its royal treasury, like this gilded kacip, exemplify this wealth. The kacip found in a pekinangan is used to cut betel nuts. Ordinary people use tools made of wood or bamboo, iron or brass. The characteristics of the kacip vary from one region to another. This one reflects the cultural influence exerted by Bali on Lombok. Animal forms, seen here in the horse, were the most popular. The decoration often included a banji, a motif not dissimilar to the swastika.

˅ 118. Cloth used as currency (*kampua*)

19th century
Buton, South-East Sulawesi
Cotton, 14.7 x 14.4 cm
MNI 2101/9327

Introduced as a form of payment in the kingdom of Wolio (or Buton, 1327–1541) to eliminate the disadvantages of bartering, the *kampua* was a piece of cotton. Around 1332, the queen Wa Kaa Kaa, the kingdom's first female ruler, began to use her clothes as currency. She cut up part of her clothes into rectangles approximately 17 cm long and gave them to inhabitants in exchange for provisions. Under the kings and sultans that followed her, the *kampua* was woven under the watchful eye of the *Bonto Ogena* (first minister), with the palace deciding the style and colour of the cloth each year. The distribution of *kampua* extended beyond the island as far as Maluku and Papua. Concerned by the growing impact of this currency in trading relations, the Dutch bought it up in quantities in order to impose the florin in its place.

141

< 119. Model of a *pinisi*

19th century
Ujung Pandang, South Sulawesi
Wood, cloth, 96 x 145 x 27 cm
MNI 663

This type of ship was used by the Bugis for commercial transportation. It originally had two masts and seven sails. During the 19th century it was influenced by European schooners.

˅ 120. *Lancang kuning*

Asahan, North Sumatra
Wood, paint, 60 x 217 x 42 cm
MNI 828

Lancang kuning means 'yellow boat'. It is the model of a very popular royal boat in Sumatra of Islamic influence, small versions of which were used to transport offerings.

> 121. Prow ornament

Langkat, North Sumatra
Wood, 98 x 68 x 30 cm
MNI 760

This figurehead is a combination of a dragon and an elephant. The form is similar to that of the *makara* in a Hindu-Buddhist temple. It was probably used on a royal boat with the purpose of chasing away evil spirits.

^ 122. Prow ornament

19th century
North Papua
Wood, 63 x 55 cm
MNI 7028

In the form of a rooster or dragon, this ornament resembles the decoration in Maluku, from where the inhabitants sailed to the north coast of Papua in the 16th century.

> 123. Prow ornament

Wood
Biak, Papua
154 x 26 x 14 cm
MNI 7029

Like more complicated prow ornaments, this object is not meant to be purely decorative; these ornaments also served as a symbol of a clan or to ward off danger.

‹ 124. Prow or stern ornament

North Papua
Wood, 50 x 97.5 x 35 cm
MNI 18533 a

The outer face is decorated with
geometric motifs in black and white.

> 125. Prow ornament

Tanimbar Islands, Maluku
Mahogany, 9 x 120 x 48 cm
MNI 14306

This prow ornament belonged to a
kora-kora boat from the Tanimbar
Islands, which was used for coastal
navigation between the islands. The
decoration consists of 'ancestor eyes',
a rooster (a symbol of nobility) and
fish (which symbolise food).

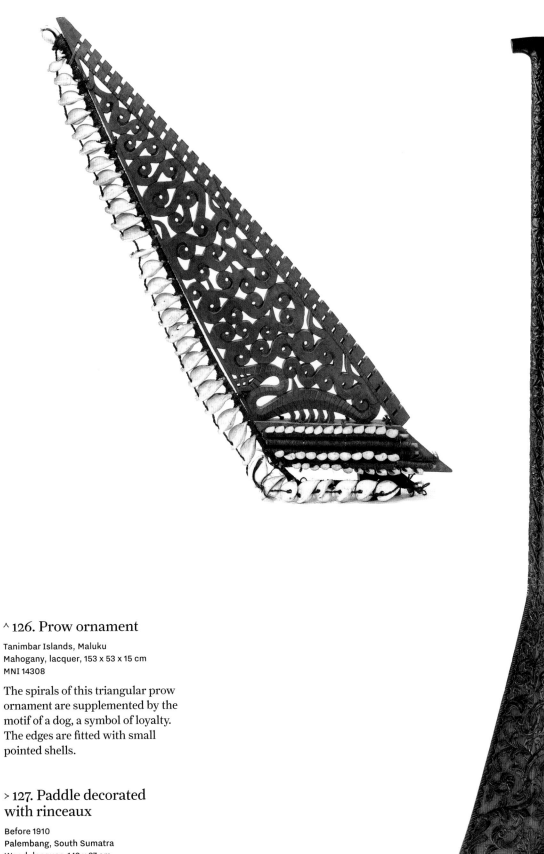

^ 126. Prow ornament

Tanimbar Islands, Maluku
Mahogany, lacquer, 153 x 53 x 15 cm
MNI 14308

The spirals of this triangular prow
ornament are supplemented by the
motif of a dog, a symbol of loyalty.
The edges are fitted with small
pointed shells.

> 127. Paddle decorated
with rinceaux

Before 1910
Palembang, South Sumatra
Wood, lacquer, 142 x 27 cm
MNI 14473

ᵛ 128. Model of a *tongkonan*

Toraja, South Sulawesi
Wood, bamboo, 117 x 120 x 36 cm
MNI 16742

With its boat-shaped roof, the *tongkonan* is the symbol of the universe. The first Toraja are said to have arrived in the region during a storm and to have used their boats as protection. *Tongkonan* were generally built on a hill for protection from enemy attack.

ⅴ **129. Model of a raft house**

Palembang, South Sumatra
Wood, bamboo, palm leaves,
58 x 71 x 58 cm
MNI 1056 a

Raft houses were the earliest form of
habitation in Palembang. Their origin
probably goes back to the kingdom of
Sriwijaya (c. 500–1377).

149

< 130. Fish trap (*bubu*)

Papua
Wood, 120 x 22 x 25 cm
MNI 26825

The use of fish traps is widespread across the archipelago. Bait is placed inside and the trap set in a water course for several days.

ᵛ 131. Harpoon (*ciruk*)

Karawang, West Java
Bamboo, iron, 68 x 23 x 4.5 cm
MNI 5143

In simple and notched form, harpoons were used for fishing in deep water.

˅ 132. Fishing net

Kalimantan
Thread, metal, 62 x 75 cm
MNI 10046

Coastal inhabitants spread their
nets on the water, then pulled
them in to catch small fish.

˅ 133. Fish trap

Sulawesi
Bamboo, 24 x 16 cm
MNI 16953

This simple type of bamboo basket
was baited to attract fish inside.

< 134. Man's shell necklace

Tanimbar Islands, Maluku
Wood, shells, fibres, 3 x 21 x 19 cm
MNI 3094

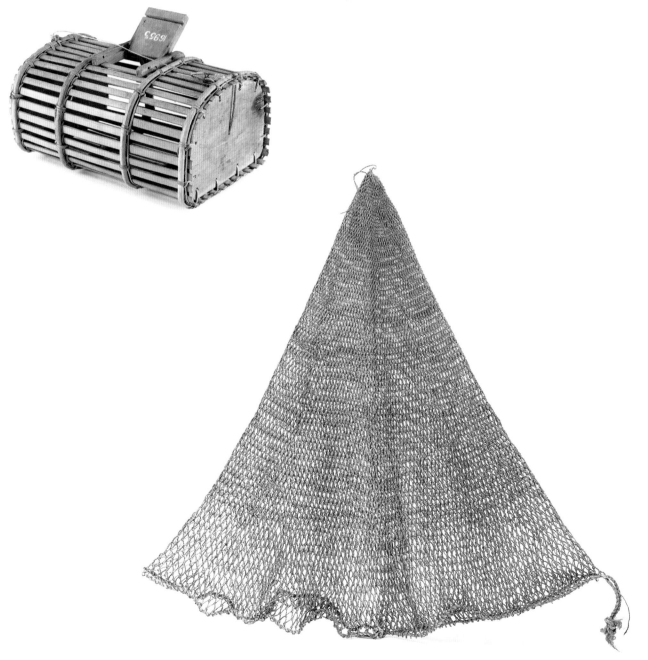

^ 135. *Tampan*

Lampung
Cotton, 73 x 72 cm
MNI 26358

The length of a *tampan* or *nampan* was
generally less than a metre. This type
of cloth was used to cover engagement
or wedding presents.

152

^ 136. Batik from *tok wi*

Pekalongan, Central Java
Cotton, 100 x 105 cm
MNI 2404

This cloth was used to cover the altar in temples and Chinese homes. Originally this form of batik was made of embroidered silk but the Indonesians made it a batik, demonstrating the cultural mix that existed. The motifs are of clouds, flowers and phoenixes. The cloths are found in coastal villages north of Java.

153

< 137. *Lokcan* batik

Lasem, Central Java
Cotton, 45.5 x 12 cm
MNI 29040

The *lokcan* is a silk batik with decorative motifs very similar to those on the north coast of Java (see cat. 136).

< 138. *Koffo* cloth

Sangir and Talaud Islands, North Sulawesi
Plantain fibres, cotton thread, 123 x 29.5 cm
MNI 27102

Made from plantain fibres (an indigenous tree on the Sangir and Talaud Islands) using the *songket* technique and cotton thread, this cloth is relatively firm and water-resistant. For this reason it was used as partitions in house interiors.

∧139. Patchwork batik

Before 1933
Cirebon, West Java
Cotton, 206 x 107 cm
MNI 20457

The 156 triangles of which this article consists attests the cultural diversity of Cirebon society: motifs of Chinese and Indian origin are mixed with traditional motifs of Java.

∧140. *Tapis* cloth

Before 1937
Krui, Lampung
Cotton, mirrors, silk thread, 107 x 62 cm
MNI 21597

This *tapis* cloth is decorated with boats, human figures, a tree of life and a post with buffalo horns, all referring to the cycle of human life (see also cat. 174).

155

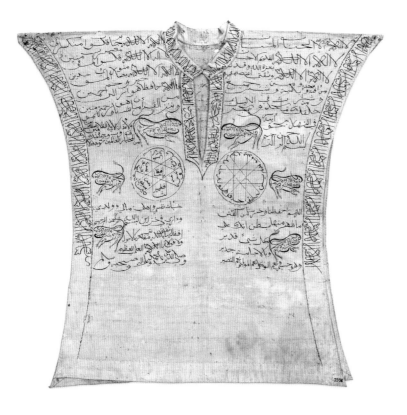

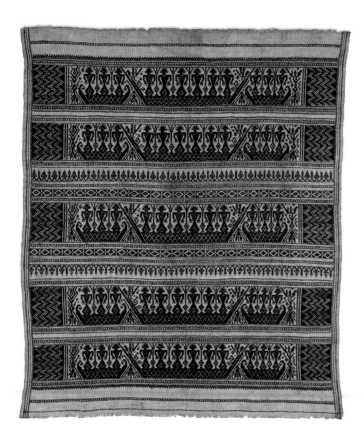

< 141. Shirt

Rote Island, East Nusa Tenggara
Cotton, 65 x 68 cm
MNI 3376

Decorated with Arabic calligraphy, this shirt is supposed to protect its wearer and give him strength.

< 142. *Tampan*

Lampung
Cotton, 74 x 70 cm
MNI 24316

This cloth is made by the Pemingirr, an ethnic group living in Lampung province. It is decorated with female figures in traditional costume standing on boats. These cloths were used by all social classes during ceremonies that accompany rites of passage.

> 143. *Tampan*

Lampung
Cotton, 53 x 53 cm
MNI 22241

The motif on this cloth is a fusion of South-East Asian, Hindu-Buddhist and Islamic influences. The triangular sailing boat and tree of life symbolise unity and strength. The red parasol indicates great spirituality. The two-pointed sword on the banners at either end of the boat is the Prophet's *zulfiqar*. The absence of human figures suggests that the cloth was made during the Islamic period.

22241.

157

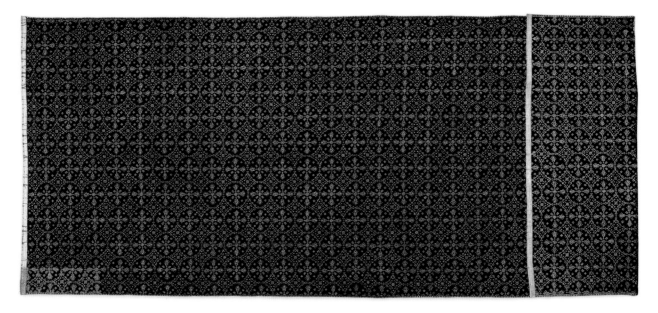

∧ 144. *Palepai*

Cotton, 342 x 62.5 cm
Ruwa Jurai, Lampung Provincial Museum, 3501

Cloths from Lampung are distinguished by their central motif: the boat symbolising the journey of human life. This type is a *palepai*, which were hung on the wall during rites of passage. Only village chiefs had the right to the *palepai* and only the eldest son could inherit it.

∧ 145. *Jlamprang*

Surakarta, Central Java
Cotton, 286 x 108 cm
MNI 23102

Jlamprang is a typical batik motif from Pekalongan based on the *patola* motif that was brought to the island by Indian migrants from Gujarat.

^ 146. Sarong

Ende, East Nusa Tenggara
Thread, 152 x 58 cm
MNI 20522

The motif on the cloth is of women playing the harp. The handpicked cotton was spun and woven using the *ikat* technique. Traditional colourants were used to create a motif clearly influenced by European culture. This sarong was worn during ceremonies.

159

> 147–148. Ceremonial outfit

South Sulawesi
Jacket: 89 x 30 cm; skirt: 96 x 97 cm

The manufacture of barkcloth has experienced a revival in the last ten years. The stylised motif of the buffalo on this costume is a sign of nobility and prosperity. This outfit is similar to the Portuguese style of female clothing in the 16th century.

> 149. Batik

1960s
Indramayu, West Java
Primissima cotton, natural colourant, 218 x 105 cm
MNI 14602

The *banji tepak* motif is typical of the Indramayu region. *Tepak* means 'small box' and *banji* is a word borrowed from Chinese that means 'multiple blessings'. These blessings are here given material form by the bountiful nature of the land and sea: fish, prawns, birds, domestic animals, rice paddies and plants.

> 150. Batik

1960s (?)
Indramayu, West Java
Primissima cotton, natural colourants, 239 x 105 cm
MNI 14603

The motif of this cloth is the *iwak etong* inspired by the sea and its bounty: prawns, crabs and sea plants.

160

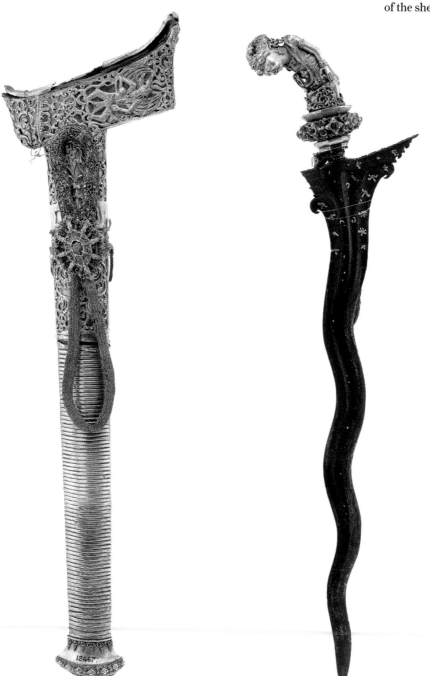

<151. Kris

Gowa, South Sulawesi
Gold, 50 x 38 x 9 cm
MNI 12467/E 591

A kris covered entirely with gold can only be a weapon used by the high nobility. In the Bugis ethnic group, a golden kris is called a *tatarepang*. Kris from the Sulawesi region are distinguished by a bulge at the end of the sheath.

∨ 152. Ceremonial container

Before 1865
Benjamarsin, South Kalimantan
Silver, gold, crystal, 65 x 40 cm
MNI 2559b/E.369

This container in the form of a scaly
dragon was used to hold betel nuts.

∨ 153. Container

18th–19th century
Lombok, West Nusa Tenggara
Silver, gold, 16.5 x 11 cm
MNI 7988/E.1026

In the form of a scaly fish, this vessel
was used to hold tobacco or betel nuts.

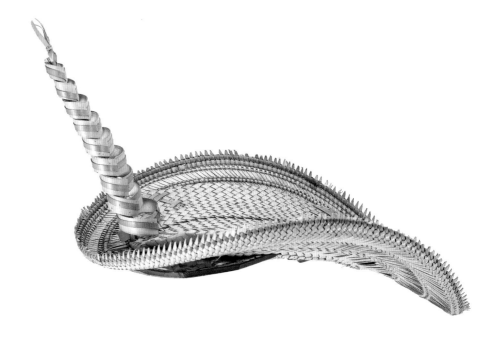

ˇ **154. Container in the form of a *nagamina***

19th century
South Bali
Wood, 25.5 x 38.5 x 24.5 cm
MNI 20828

The *nagamina* is a mythical creature that variously combines a dragon, an eagle and a fish. It was believed that this animal attracted fish and protected fishermen.

ˇ **155. *Sasando***

Rote Island, East Nusa Tenggara
Bamboo, *lontar* leaves, thread,
41 x 66 x 56 cm
MNI 3393

A musical instrument made from palm leaves and bamboo, the *sasando* might have as many as 22 strings. Its player sits cross-legged and plucks the strings. The palm tree is fundamental to the life of the inhabitants of Rote and its neighbouring island Savu. Its juice provides nourishment and its leaves are used to make baskets, roofs and musical instruments.

^ **156. Hat (*ti'i langga*)**

Rote Island, East Nusa Tenggara
lontar leaves, 35 x 57 x 39 cm
MNI 28681

This traditional hat is worn on Rote Island during cultural events, such as the playing of the *sasando* (see cat. 155).

164

ˇ 157. Water spout

Before 1910
Cirebon, West Java
Copper, 41 x 28 cm
MNI 14313

This water spout in the shape of a
dragon was used in Sunyaragi, the
water palace of the sultans of Cirebon.

165

23793.
INI GAMBAR RADJA
van Lombok Ratoe
Agoeng Agoeng Gede
Ngoerah Karang Assem
Bikinan Gede Bem

< 158. Statuette of the king of Lombok

Cakranegara, Lombok, West Nusa Tenggara
Wood, 45.3 x 19.6 x 17.6 cm
MNI 23793

In 1894 the kingdom of Cakranegara in Lombok was attacked by a Dutch military expedition. The kingdom was devastated and the royal treasures removed to Batavia. The king, Anak Agung Gde Ngurah Karangasem, was imprisoned in Tanah Abang (near Batavia), where he died in 1895.

> 159. Crown

Lombok, West Nusa Tenggara
Velvet, cotton, gold, 5 x 20 cm
MNI 8016/E 1101

This gold crown lined with red velvet
was worn by nobles in the kingdom
of Lombok.

> 160. *Kupiah meukeutop*

Aceh
Cotton, gemstones, gold, 13 x 18.5 cm
MNI 754/E 61

The *kupiah meukeutop*, the traditional
hat worn in Aceh, was used in
ceremonies. For the inhabitants of
Aceh, the hat is associated with their
national hero Teuku Umar (1854–1899),
who was never without one.

∨ 161. Bracelet

Before 1905
Gowa, South Sulawesi
Gold, 1 x 9 x 7 cm
MNI 17476/E 5231

This pair of bracelets in the form of
dragons is a vestige of the sultanate
of Gowa.

^ 162. Crown

Java
Gold, 12.5 x 18 cm
MNI 1065

The Islamic characteristics of this crown have affinities with those of a crown of the kingdom of Surakarta.

> 163. Crown

Klungkung, Bali
Bamboo, gold, silver, precious stones,
27.5 x 21.5 cm
MNI 14897/E.827

This crown was probably worn by a female dancer during ceremonies at the royal palace in Klungkung.

∧ 164. Necklace

Before 1950
Kutai, East Kalimantan
Gold, 1.5 x 25 x 26 cm
MNI 27788/E.1273

This gold necklace decorated with 24 fish pendants was used as a talisman by the king of Kutai.

> 165. Necklace

Before 1894
Lombok, West Nusa Tenggara
Gold, 1.5 x 33 cm
MNI 2364/144 (LB-5C)

Both a jewel and a talisman, this necklace adorned with Chinese coins and various sea creatures exemplifies the cultural influence of the Chinese, Indian and Arab immigrants locally.

^ 166. Plate

Early 20th century
Palembang, South Sumatra
Wood, lacquer, 3.5 x 47 cm
MNI 24303

The use of lacquer on this plate is Chinese in origin. It spread to Sumatra as a result of the development of relations between China and the kingdom of Sriwijaya beginning in the 7th century. Although the goldfish motif dates the plate to the start of the 20th century, the motifs as a whole are of Chinese inspiration.

> 167–170. Cloth used as currency

19th century
Buton, South-East Sulawesi
13002: Cotton, 14.8 x 14.3 cm
13703: Cotton, 14 x 13 cm
13706: Cotton, 14 x 14.5 cm
25759: Cotton, 14 x 15 cm
NMI 13002, 13703, 13076 and 25759

This small, rectangular piece of cloth was used as a form of currency. Its style, which was altered at regular intervals to prevent forgery, was determined by the prime minister.

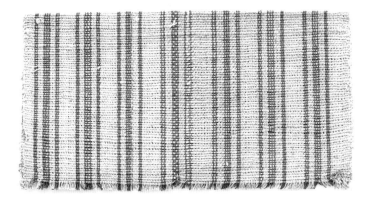

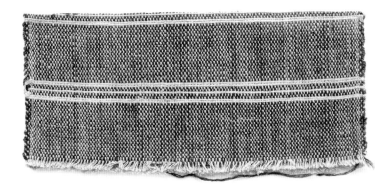

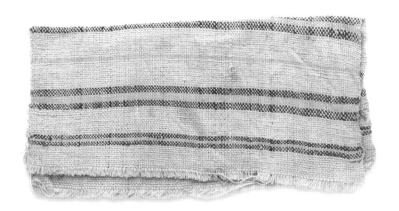

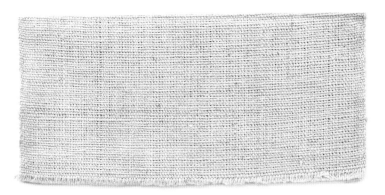

> 171. Shell

Maluku
24 x 10 x 14 cm
MNI VI D 11

This shell comes from a large gastropod that lives in shallow water. In Maluku, shells like these are used as wind instruments.

⌄172. Man's shirt (*tampan*)

Kroe Marga Lima, Sumatra
Cotton, 58 x 40 x 130 cm
MNI 4140

> 173. Sarong made from batik

Madura, East Java
Cotton, 192 x 110 cm
MNI 27274

The sarong is decorated with several motifs, in particular the *tumpal* (triangles), the *kopi pecah* (coffee beans) and the *jamblang* (a woven bamboo basket).

> 174. *Tapis inuh*

Cotton, silk thread, 129 x 66 cm
MNI 20439

In Lampung, *tapis* is the name of both a weaving technique and the resulting cloth. A *tapis* is worn as a sarong. One of the highest quality cloths in Lampung, this cloth is woven using the Indonesian *ikat* technique and enhanced with the Chinese tradition of silk embroidery. The central motif of the squid was a prerogative of the nobility.

173

^ **175. Woven batik**

Pekalongan, Central Java
Cotton, 210 x 81 cm
MNI 29008

The European features of this cloth
– flat tints with repeated motifs –
suggest that it was made for the
colonial market.

> **176. Batik stamp**

Before 1960
Japara, Central Java
Wood, 2 x 23 x 12 cm
MNI 27524b

The use from 1840 of a wooden stamp
enabled craftsmen to speed up the
manufacture of batik and thus offer
it at lower prices. The motif here is
of Gathotkaca, a popular figure from
shadow theatre.

> **177. Batik map of Indonesia**

Pekalongan, Central Java
Cotton, 160 x 52 cm
MNI 26533

In the multicultural city of Pekalongan, batik cloths were often supports
for paintings. In this case, the image
is a map of Indonesia during the time
of Dutch colonisation.

177

<< 178. *Hinggi kombu*

Cotton, natural dyes
East Sumba, East Nusa Tenggara
284 x 182 cm
MNI 18828

Traditionally worn by the men of
Sumba during rites of passage,
the colour and motifs of the *ikat*,
associated with the wearer's age,
were related to his social status.

<< 179. *Tampan*

Late 19th century
Krui, Lampung
Cotton, 119 x 56.5 cm
MNI 23356

This rare *tampan* cloth is decorated
with the tree of life, a motif known
since Austronesian prehistory. The
tree of life is believed to be the place
where the ancestors live and where
it is possible to enter into communication with them.

^ 180. Batik

Pekalongan, Central Java
Cotton, 206 x 106 cm
MNI 27270

Batiks made on the coast of Java were
affected by many different influences.
Here a motif of Dutch origin is mixed
with traditional motifs of flowers and
birds.

∧ 181. Batik sarong

Lasem, Central Java
Cotton, 164 x 61 cm
MNI 29015

Red is referred to as 'chicken blood' and is obtained a mixture of roots and the local water, which has particular mineral properties. There are two types of motifs on Lasem batiks: those influenced by Chinese culture, and native motifs like the triangular *tumpal*, bamboo shoots and animals.

179

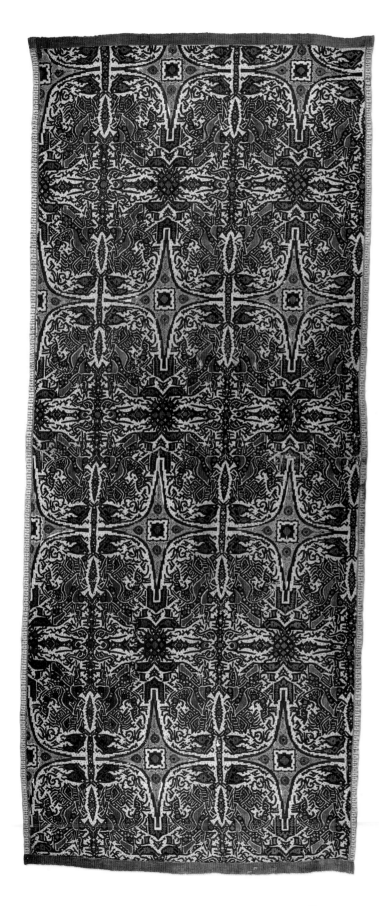

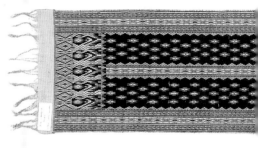

< **184. Ceremonial cloth**
(*salempuri*)

Lampung, Sumatra
Cotton, 277 x 118 cm
MNI 19098

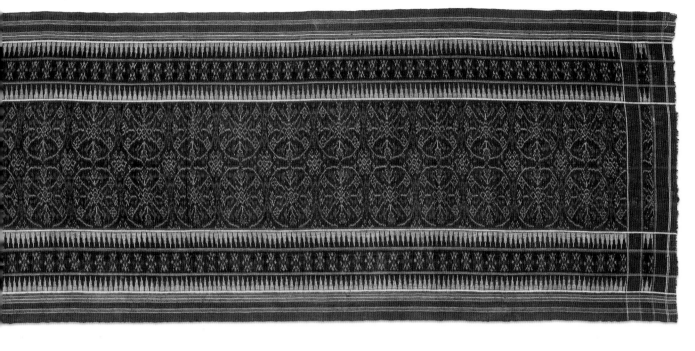

∧185. Cloth (*cepuk*)

Nusa Penida, Bali
Silk, gold thread, 260 x 79 cm
MNI 18834

Cepuk means 'meeting (with the divine powers)'. Only women from the Balinese royal family were allowed to wear this cloth made with a combination of the *ikat* and *songket* techniques. The colours were applied using a bamboo brush (the *coletan* technique).

∧187. Shawl

Bira, South Sulawesi
Cotton, 237 x 28 cm
MNI 16896 a

Bira is known for its woven goods. Weaving was intrinsic to maritime trade with other regions in the archipelago.

181

^183. Man's shirt

Lampung, Sumatra
Cotton, shells, 142 x 30 cm
MNI 20808

⌄182. Table runner

Before 1933
Jailolo, North Maluku
Cotton, 92 x 32 cm
MNI 20483

This ornamental table runner is decorated with *tassels* and *lozenge* motifs.

>186. Man's jacket

Kalimantan
Heavy traditional cotton, 39 x 33 cm
MNI 18579

The complex use of red, white, yellow and black in this striped *ikat* jacket is distinctive of the Dayak.

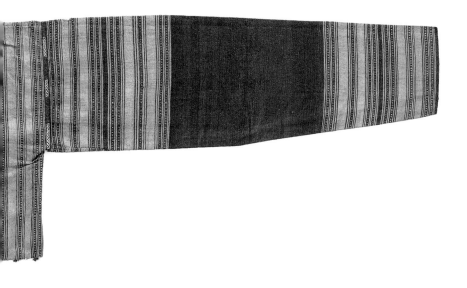

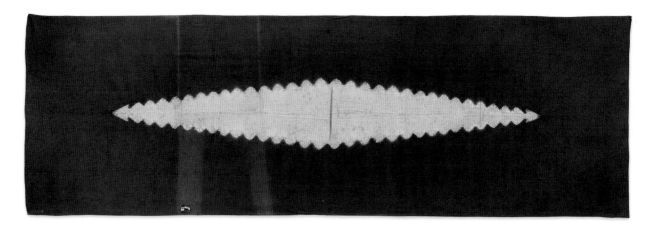

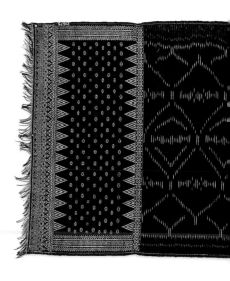

< 188. *Jumputan* cloth

Palembang
Silk, gold thread, 261 x 39 cm
MNI 12912

Jumputan is derived from the word *jumput*, which means tie-dyeing.

< 189. *Kembangan*

Surakarta, Central Java
Cotton, 148 x 47.5 cm
MNI 14438

Customarily used like *kemben* to cover the torso, this cloth is often decorated with a pale apotropaic lozenge surrounded by a dark colour.

∨ 190. Shawl

Pasemah, Palembang, South Sumatra
Cotton, gold thread, silver thread, 174 x 43 cm
MNI 21675

In Palembang, the *songket* is traditionally a wedding gift that allows the husband to provide his wife with the garment she will wear at traditional ceremonies. Until the 19th century, *songket* were worn exclusively by women.

∧ 191. *Sireuw*

Papua
Cotton, beads, fibres, 66 x 64 cm
MNI 27053

The *sireuw* is a five-sided skirt worn by noblewomen during traditional ceremonies, in particular for dancing. It is made with beads and cotton fibres to form geometric motifs. The lower part is decorated with pieces of white and red cloth. The *sireuw* was part of a dowry and a form of compensation to settle a dispute.

185

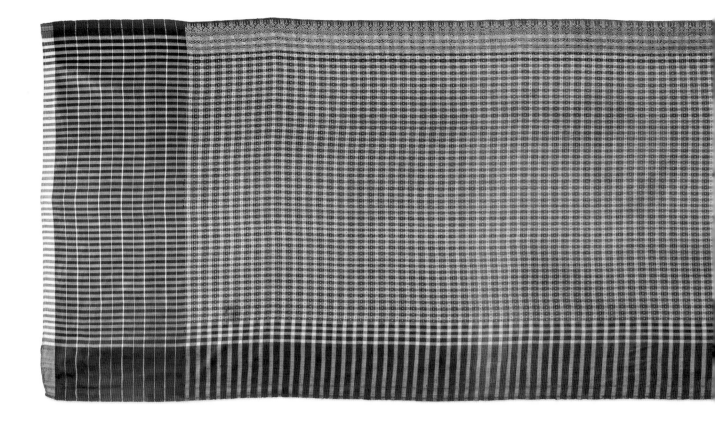

∧ 192. Sarong

South Sulawesi
Silk, 226 x 89 cm
MNI 26553

The cloths of the Bugis are decorated
with contrasts of pale colours. If, like
this one, they are checked, the young
woman who wears it is unmarried.
Here the purple at the bottom of the
cloth has been chosen to relate to
the glint of gold thread in the *tumpal*
motif in the upper section of the
sarong.

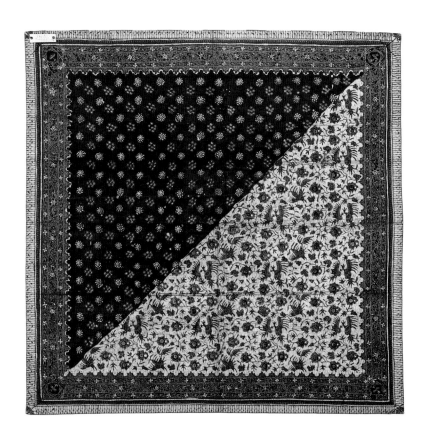

∧ 193. Man's headdress

Baturaja, Palembang, South Sumatra
Cotton, 103 x 102 cm
MNI 23680

The birds on this batik were inspired by the *hong* (the mythical Chinese bird, see cat. 93) and floral motifs.

∧ 194. Bandolier

Timor, East Nusa Tenggara
Cotton, 170 x 11 cm
MNI 23650

This bandolier was worn by the *meo*, elite Timorese troops who protected the border.

< 195. *Patola*

Yogyakarta
Silk, 205 x 80.5 cm
MNI 23719 R

Indonesian craftsmen took up the
patola motif on cloths from Gujarat
(India), which were used as a form
of payment for spices.

> 196. *Songket*

Before 1936
Singaraja, Bali
Silk, gold thread, 181 x 155 cm
MNI 21656

The *songket* is a predominantly red,
handwoven cloth in the family of
brocades in which gold or silver
thread is woven transversally.

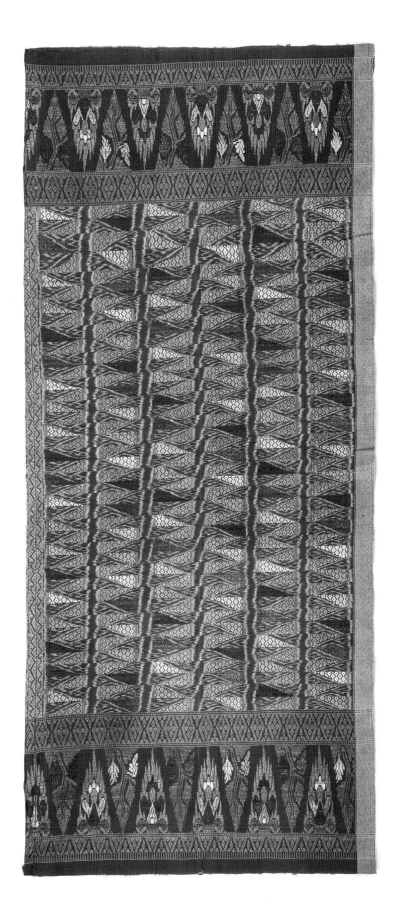

190

< 197. Blouse

Takengon, Aceh
Cotton, 53 x 47 cm
MNI 23103

Sewn by men and worn by women, this garment is decorated with geometric motifs typical of the Gayo ethnic group in Aceh. The circle represents the cycle of life, however, an Islamic interpretation allows it to be associated with the moon.

^ 198. Jacket

Menggala, Lampung, South Sumatra
Cotton, velvet, gold thread, 47 x 40 cm
MNI 22388

The *songket* (brocade) technique is also used in the weaving of men's clothing. This jacket is influenced by the styles of the Middle East.

< 199. Skirt

Dayak Taman, West Kalimantan
Beads, cotton, 43 x 43 cm
MNI 28461

In Kalimantan, beads are highly valued jewels. Known since prehistory, beads arrived in Kalimantan through commercial relations, mainly with China.

191

< 200. *Sidomukti* batik

Solo, Central Java
Cotton, 422 x 107 cm
MNI 28166

Made with natural brown pigments,
this batik from Solo gets its name
from two words: *sido* ('who becomes')
and *mukti* ('noble, prosperous').

∧ 201. *Saput*

Before 1940
South Bali
Silk, 141.5 x 122.5 cm
MNI 23939

The *saput* is a skirt worn by men.
This one was made using the *songket*
technique.

> 202. Model of a *rumah gadang* ('big house')

Minangkabau, West Sumatra
Wood, palm tree fibres, rattan, 91 x 122 x 80 cm
NMI 278b

This is the house of the local chief of Padang. The form of the roof is called 'elephant feeding her calf'. *Rumah gadang* are inhabited by great matrilinear families.

˅ 203. Model of a house of a Muslim religious expert (*kyai*)

Kudus, Central Java
Wood, 100 x 121 x 125 cm
MNI 1319

This type of house is distinguished by the form of its roof (*joglo*) and the fact that there are no interior walls, the spaces being divided by decorative carved screens.

194

› 204. Model of a house with a *limasan* roof

Karawang, West Java
Wood, bamboo, 55 x 54 x 60 cm
MNI 4755

This house has a *limasan* (pyramidal) roof and a raised wooden floor. A covered terrace in front has half-walls made from woven bamboo. The entrance is reached via the steps to the left and right of the terrace.

ᵛ 205. Model of a *dayak* stilt house

19th century
Pontianak, East Kalimantan
Wood, 83 x 93 x 67 cm
MNI 26638

The house stands on ten stilts carved at the top with flower petals. The walls are usually made from shingles of hardwood.

> **206. Model of a traditional *batak* house**

19th century
North Sumatra
Wood, palm tree fibre, rattan, 106 x 97 x 74 cm
MNI 23610

Aligned north-south, *batak* houses are embellished with *gorga*, sculptures that ward off evil.

v **207. Model of a traditional house**

19th century
Kerinci, Jambi
Wood, shingles, *lontar* leaves, 53 x 71 x 46 cm
MNI 23453

The *Umoh Laheik* or *Umoh Panja*, the traditional house in Kerinci, has no fixed foundations, just a few stones piled up to support the posts. It is made without the use of nails, just dowels and palm-fibre ropes. Palm leaves are also used to cover the house. The lower section of the house (*umin*) was used to store farm tools or to stable animals. Daylight enters the loft through a small window close to the crest of the roof.

> 208. Model of a mosque

19th century
East Java
Wood, bamboo, 145 x 82 x 122 cm

Crowned by a stupa, this mosque is
evidently in Hindu-Buddhist style.

∨ 209. Tray

19th century
Cirebon, West Java
Wood, 5 x 40 x 26 cm
MNI 21207

Cirebon is a coastal town so it is
unsurprising that the goods that it
produces are often blue in colour
and decorated with motifs of fish
and clouds. The latter are part of
the *mega mendung* absorbed from
Chinese culture.

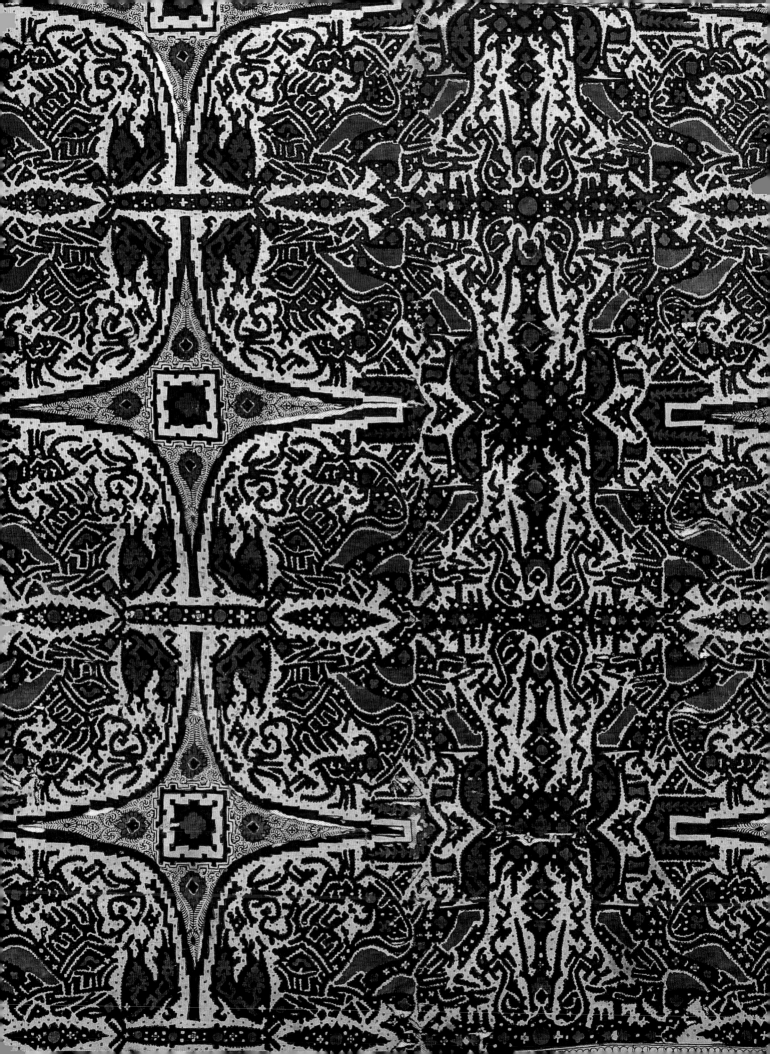

GOVERNMENT ANNOUNCEMENT
ON THE TERRITORIAL SEA
OF THE REPUBLIC OF INDONESIA

The Council of Ministers at its meeting on Friday, 13 December 1957 to discuss the territorial sea of the Republic of Indonesia.

The geographic state of Indonesia as an archipelagic entity composed of (thousands) of islands with its own unique characteristics.

In order to maintain the territorial integrity of the nation and to protect the property of the State, all islands and the sea between them must be considered as a single unity.

The delimitation of the territorial sea as defined in the Territorial Waters and Maritime Zones Ordinance of 1939 (Stb 1939 n ° 442). Article 1, paragraph 1 no longer conforms with the considerations above, on account of the division of Indonesian land into separate zones within its own territory.

On the basis of these considerations, the Government states that all the waters around the islands or the parts of islands belonging to the Republic of Indonesia, irrespective of their breadth or size, naturally belong to its territory and therefore constitute an integral part of the internal or national waters that are subject to the absolute sovereignty of Indonesia. The peaceful passage of foreign ships through these waters is guaranteed for so long and in so far as it is not contrary to the sovereignty of the Indonesian State, or prejudicial to security.

The delimitation of the territorial sea, extending a distance of 12 nautical miles, shall be measured in straight lines from the most extreme points of the islands of the Republic of Indonesia.

The provisions mentioned above will be regulated in the form of a law as soon as possible.

The positions of this Government will be discussed at an international conference on the Law of the Sea to be held in February 1958 in Geneva.

Jakarta, 13 December 1957
Prime Minister
Signed. H. Djuanda

BIBLIOGRAPHY

Abdullah, Taufik
  2002. 'Pemikiran Islam di Nusantara Dalam Perspektif Sejarah'. *Makalah diskusi peluncuran buku Ensiklopedi Tematos Dunia Islam*. Jakarta, 5 September.
Abdullah, Taufik & A.B. Lapian (ed.)
  2012. *Indonesia dalam Arus Sejarah (Jilid IV)*. Bandung: Ichtiar Baru.
Aboebakar, H.
  1955. *Sedjarah Mesjid I dan II, dan Amal Ibadah di Dalamnya*. Jakarta: NV. Viss and Co.
Adhyatman, Sumarah
  1982. *Keramik Kuno yang Ditemukan di Indonesia, Berbagai Penggunaan dan Tempat Asal*. Jakarta-Indonesia: the Ceramic Society of Indonesia.
A.H.H.
  1935. 'Shipping in the Netherlands Indië', *Netherlands Indië*, vol. III, no. IV, pp. 70-84.
Alfian, T. Ibrahim
  1986. *Mata Uang Emas Kerajaan-Kerajaan di Aceh. Daerah Istimewa Aceh*: Depdikbud, Proyek Pengembangan Permuseuman.
Alimuddin, M.R.
  2005. *Orang Mandar Orang Laut*. Jakarta: Kepustakaan Populer Gramedia.
Ambary, Hasan Muarif
  1984. 'Penelitian Arkeologi yang Baru di Sumatera'. *Amerta*. Jakarta: Pusat Penelitian Arkeologi Nasional.
  1990. 'Peranan Beberapa Bandar Utama di Sumatera Abad ke-7-16 Masehi'. In: *Jalur Jalan Melalui Lautan'. Saraswati-Kalpataru, Majalah Arkeologi*, no. 9. Depdikbud.
  1996. 'Kaligrafi Islam Indonesia: Dimensi dan Signifikansinya dari Kajian Arkeologi'. *Jurnal Arkeologi Indonesia*, no. 2. Jakarta: Ikatan Ahli Arkeologi Indonesia.
  1996. 'Makam-Makam Islam di Aceh'. *Aspek-Aspek Arkeologi Indonesia*, no. 19. Jakarta: Pusat Penelitian Arkeologi Nasional.
  1997. 'Tinggalan Arkeologi Samudra Pasai'. *Pasai Kota Pelabuhan Jalan Sutra: Kumpulan Makalah Diskusi*. Jakarta: Depdikbud.
  1998. *Menemukan Peradaban: Arkeologi dan Islam di Indonesia*. Jakarta: Pusat Penelitian Arkeologi Nasional.
Ammarell, G.
  2008. *Navigasi Bugis*. Makassar: Hasanuddin University Press.
Andaya, Leonard Y.
  1995. 'The Bugis-Makassar Diasporas'. *Journal of the Malaysian Branch of the Royal Asiatic Society*, vol. 68 (1995), 1.
Anonymous, 2017. *Ekspedisi Ke Jepang, Kapal Spirit of Majapahit Berlabuh Di Manila*, http://beritatrans.com/2016/05/30/ekspedisi-ke-jepang-kapal-spirit-of-majapahit-berlabuh-di-manila. (assessed 25 March 2017).
Anonymous, 2017. *Kapal Ekspedisi Majapahit Siap Berlabuh di Tokyo*, http://bisnis.liputan6.com/read/2560287/kapal-ekspedisi-majapahit-siap-berlabuh-di-tokyo. (assessed 25 March 2017).

Asnan, Gusti
  2011. *Penetrasi Lewat Laut: Kapal-kapal Jepang di Indonesia Sebelum Tahun 1942*. Yogyakarta: Ombak.
Astuti, Meta Sekar Puji
  2008. *Apakah Mereka Mata-Mata? Orang-orang Jepang di Indonesia (186-1942)*. Yogyakarta: Ombak.
Atmodjo, Junus Satrio (ed.)
  1999. *Masjid Kuno Indonesia*. Jakarta: Depdikbud, Direktorat Perlindungan dan Pembinaan Peninggalan Sejarah dan Purbakala.
Azra, Azyumardi
  1999. *Renaisans Islam Asia Tenggara: Sejarah Wacana dan Kekuasaan*. Jakarta: Rosda.
Barnes, Ruth (ed.)
  2005. *Textiles in Indian Ocean Societies*. New York: Routledge Curzon.
Bart, Bernhard
  2006. *Revitalisasi Songket Lama Minangkabau*, Padang: Studio Songket ErikaRianti.
Bock, Carl
  1882. *The Head Hunters of Borneo: A Narrative of Travel Up the Mahakkam and Down the Barito*; Also: *Journeyings in Sumatra*. London: S. Low, Marston, Searl & Rivington.
Boxer, C.R.
  1965. *The Dutch Seaborne Empire, 1600-1800*. London: Hutchinson.
Brandes, J.L.A.
  1886. 'Een Nagari Opschrift: Gevonden tusschen Kalasan en Prambanan', *TBG* 31, pp. 240-250.
Brinkgreve, Francine, Pauline Lunsingh Scheurleer & David Stuart-Fox
  2010. *Kemegahan Emas di Museum Nasional Indonesia*. Jakarta: Duta Tiga Perkasa.
Brinkgreve, Francine & Retno Sulistianingsih (ed.)
  2009. *Sumatera: Crossroads of Culture*. Leiden: KITLV Press.
Brownrigg, Henry
  1992. *Betel Cutter from the Samuel Eilenberg Collection: From the Samuel Eilenberg Collection*. Thames & Hudson.
Budi Utomo, Bambang
  2012. *Atlas Sejarah Indonesia: Masa Klasik (Hindu-Buddha)*, Endjat Djaenuderadjat (ed.), Direktorat Jenderal Sejarah dan Purbakala, Kementerian Pendidikan dan Kebudayaan, Jakarta: PT Kharisma Ilmu.
  2016. *Pengaruh Kebudayaan India dalam Bentuk Arca di Sumatra*. Jakarta: Yayasan Pustaka Obor Indonesia.
  2016. *Warisan Bahari Indonesia*. Jakarta: Yayasan Pustaka Obor Indonesia.
Budi Utomo, Bambang et al.
  2009. *Treasures of Sumatra*. Jakarta: Museum Nasional Indonesia.
  2016. *The Maritime Legacy of Indonesia*. Jakarta: The National Museum of Indonesia.

Campo, J.N.F.M.1992. *'De Koninklijke Paket-vaart Maatschappij: Stoomvaart en staatsvorming' in de Indische archipel 1888-1914.* Hilversum: Verloren.

Caro, P.
2012. *Ekspedisi Phinisi Nusantara: Pelayaran 69 hari Mengarungi Samudera Pasifik.* Jakarta: Kompas.

Cense, A.A & H.J. Heeren
1972. *Pelajaran dan Pengaruh Kebudajaan Makassar-Bugis di Pantai Utara Australia.* Jakarta: Bhratara.

Chauduri, K.N.
1989. *Trade and Civilization in Indian Ocean: An Economic History from the Rise of Islam to 1750.* Cambridge: Cambridge University Press.

Christie, J.W.
1998. 'Javanese markets and the Asian Sea Trade Boom of the Tenth to Thirteenth Centuries A.D'. *Journal of the Social and Economic History of the Orient*, vol. 41, no. 3.
1999. 'Asian Sea Trade between the Tenth and Thriteenth Centuries and Its Impact on the States of Java and Bali'. In: H.P. Ray (ed.), *Archeology of Seafaring: The Indian Ocean in the Ancient Perion.* Delhi: Pragati.

Coedes, G.
1968. *The Indianised States of Southeast Asia.* Kuala Lumpur: University of Malaya Press.

Collins, G.E.P.
1992. *Makassar Sailing.* Singapore: Oxford University Press.

Cortesao, A. (ed.)
1944. *The Suma Oriental of Tome Pires: An Account of the East.* London: Hakluyt Society.

Cribb, Robert
2000. *Historical Atlas of Indonesia* (Honolulu: Univrsity of Hawaii Press).

Crill, Rosemary
1998. *Indian Ikats Textiles.* London: V & A Publications.

Dahlan, Juniawan
2011. *Pengamatan Gaya dan Perbandingan Figurin Terakota Manusia di Trowulan dan Relief-Relief Candi Masa Singhasari dan Majapahit.* Skripsi. Depok: FIB-UI.

Damais, L.Ch.
1955. 'Étude d'Épigraphie Indonésienne: IV. Discussion de la Date des Inscriptions', *BEFEO*, XLVII, pp. 1-290.

Danusaputro, M.
1980. *Tata Lautan Nusantara dalam Hukum dan Sejarahnya.* Bandung: Binacipta.

Davie, Don. Cannons of the Malay Archipelago. Diakses dari http://www.acant.org.au/Articles/MalayCannons.html

Degroot, Véronique
2009. *Candi Space and Landscape: A Study on the Distribution, Orientation and Spatial Organization of Central Javanese Temple Remains.* Leiden: Sidestone Press in cooperation with National Museum of Ethnology Leiden.

Departemen Pendidikan & Kebudayaan
1997. *Budaya Menginang di Daerah Irian Jaya, Maluku, dan Sulawesi.* Jakarta: Proyek Pembinaan Permuseuman.

Dick, Howard
1989. 'Japan's Economic Expansion in the Netherlands Indie between the First and Second World War'. *Journal of Southeast Asia Studies*, vol. XX, no. 2, Sep. 1989, pp. 244-272.

Disjar
1973. *Sejarah Tentara Nasional Indonesia Angkatan Laut (Periode Perang Kemerdekaan) 1945-1950.* Jakarta: Dinas Sejarah TNI-AL.

Djafar, Hasan
2002 'Kompleks Percandian Buddhis di Kawasan Batujaya, Karawang, Jawa Barat'. In: [cat. exh.] *Fajar Masa Sejarah Nusantara.* Jakarta: Museum Nasional, pp. 49-57.

Djalal, H.
1979. *Perjuangan Indonesia di bidang Hukum Laut.* Bandung: Binacipta.

Domenig, G.
2014. *Religion and Architecture in Premodern Indonesia: Studies in Spatial Anthropology.*

Duha, Nata'alui
2012. *OMO NIHA –Perahu Darat di Pulau Bergoyang.* Gunungsitoli: Museum Pusaka Nias.

Faruqi, Ismail R.
1992. *Atlas Kebudayaan Islam.* Kuala Lumpur: Dewan Bahasa dan Pustaka.

Fathurrahman, Oman
2005. 'Naskah dan Rekonstruksi Sejarah Islam Lokal: Contoh Kasus dari Minangkabau'. *Mimbar*, vol. 22, no. 3, pp. 260-268.

Flines, E.W. van Orsoy de
1937. 'De Keramische Verzameling'. *Jaarboek Koninklijk Bataviaasch Genootschap van Kunsten en Wetenschappen*, IV, 1937.
1969. *Guide to the Ceramic Collection of the Museum Pusat Jakarta.* Second edition. Djakarta: Bhratara.

Fontein, Jan, R. Soekmono & Edi Sedyawati
1990. *The Sculpture of Indonsia.* Washington: The National Gallery of Art, and New York: Harry Abrams.

Gillow, John & Nicolas Barnard
1991. *Traditional Indian Textiles.* Singapore: Thames and Hudson.

Gopinatha Rao, T.A.
1904. *Elements of Hindu Iconography*, 2 vols., Madras: The Law Printing House.

Groeneveldt, W.P.
1960. *Historical, Notes on Indonesia & Malaya Compiled from Chinese Sources.* Jakarta: Bhratara.
2009. *Nusantara Dalam Catatan Tioghoa.* Depok: Komunitas Bambu.

Gunder-Frank, Andre
1998. *Reorient: Global Economy in the Asian Age* (Berkeley-Los Angeles-London: University of California Press).

Hadi W.M., Abdul
1998. *Karya-karya Terpilih Kesusastraan Arab dan Parsi.* Modul Kuliah Pusat Pengajian Jarak Jauh, Universiti Sains Malaysia, P. Pinang, Malaysia Ibrahim Alfian (1999) *Wajah Aceh Dalam Lintasan Sejarah.* Banda Aceh: Pusat Dokumentasi dan Informasi Aceh.

Hadiwijoyo, Harun
1971. *Agama Hindu dan Agama Buddha.* Djakarta: BPK.

Hall, D.G.E.
1964. *A History of Southeast Asia.* London: Oxford University Press.

Hall, K.R
1985. *Maritime Trade and State Development in Early Southeast Asi*a (Honolulu, Hawaii: University of Hawaii Press, 1985).

Hamid, A.R.
2011. *Orang Buton: Suku Bangsa Bahari Indonesia.* Yogyakarta: Ombak.

Hamid, Ismail
1983. *Kesusasteraan Melayu Lama dari Warisan Peradaban Islam.* Petaling Jaya, Selangor: Fajar Bakti Sdn. Bhd.

Hardiati, Endang Sri & Sutrisno, M.M.
2000. *Kajian Ilmiah: Wawasan Seni dan Teknologi Terakota Indonesia.* Jakarta: Museum Nasional.

Hardiati, Endang Sri & Agus Aris Munandar (ed.)
2002. *Arca Dewa-Dewa Hindu Koleksi Museum Nasional.* Jakarta: Museum Nasional.

Hardiati, Endang Sri
2002. 'Seni Arca Masa Hindu-Buddha di Jambi'. *25 Tahun Kerjasama Pusat Penelitian Arkeologi dan Ecole Francaise d'Extreme-Orient.* Bogor: Grafika Mardi Yuana.
2008. *Sejarah Nasional Indonesia II, Zaman Kuno.* Edisi Pemutakhiran. Cetakan Kedua. Jakarta: Balai Pustaka.

Härmmerle, P. Johannes Maria
2001. *Asal Usul Masyarakat Nias: Suatu Interpretasi.* Gunungsitoli: Penerbit Yayasan Pusaka Nias.

Haryono, Timbul
2005. 'Sekilas Tentang Sejarah Perkembangan Kebudayaan Logam'. In: [cat. exh.] *Peranan Logam dalam Kehidupan Masyarakat Indonesia*, Jakarta: Museum Nasional.

Hatanaka, Kokyo
1996. *Textiles Arts of India*, San Francisco: Chronical Book.

He Li
2006. *Chinese Ceramics, the New Standard Guide, from the Asian Art Museum of San Francisco.* Singapore: Thames & Hudson.

Hirth, Friedrich & Rockhill, W.W.
1965. *Chau Ju-Kua, His Work on the Chinese and Arab Trade in the 12th – 13th centuries, titled Chu Fanchi.* Taiwan: Literature House, Ltd.

Horridge, Adrian

1981. *The Prahu: Traditional Sailing Boat of Indonesia*. Melbourne: Oxford Univeristy Press.

1986. *Sailing Craft of Asia*. Singapore: Oxford University Press.

2015. *Perahu Layar Tradisional Nusantara*. Yogyakarta: Penerbit ombak. http://Kridewaruci.com/history

Indraningsih Pangabean, J.R.

1978. 'Kerangka Penelitian Manik-manik Indonesia'. In: *Lokakarya Arkeologi 21-26 Februari*. Jakarta: Pusat Penelitian Arkeologi Nasional.

Institute dan Vlekke, Bernard M.K.

2008. Nusantara: Sejarah Indonesia. Cetakan Kedua. Jakarta: KPG, Freedom Balai Pustaka.

Iqbal, Muhammad Zafar

2006. *Kafilah Budaya: Pengaruh Persia Terhadap Kebudayaan Indonesia*. Jakarta: Citra.

Irvine, David

2005. *Leather Gods & Wooden Heroes: Java's Classical Wayang*. Singapore: Times Editions.

Ismail, Ibrahim

1989. 'Pengaruh Parsi dalam Sastra Melayu Islam'. *Journal Ulumul Qur'an*, vol. I, 1989/1440, pp. 38-44.

Joedodibroto, Rijadi

2008. *Mengenal Arsitektur Nias dalam Nias dari Masa Lalu ke Masa Depan*. Jakarta: Badan Pelestrarian Pusaka Indonesia, pp. 181-263.

Jonge, Nico De & Toos Van Dijk

1995. *Forgotten Islands of Indonesia: The Art & Culture of The Shoutheast Maluku*. Singapore: Periplus.

Jorg, C.J.A.

1983. *Oosters Porselein, Delfts Aardewerk - Wisselwerkingen*. Groningen: Kemper.

Kain Tradisional Nusantara Seribu Nuansa satu Indonesia (cat. exh.)

2016. Jakarta: Museum Nasional Indonesia.

Kalus, Ludvik

2008. 'Sumber-Sumber Epigrafi Islam di Barus'. *Barus Seribu Tahun Yang Lalu*. Jakarta: Forum Jakarta-Paris, pp. 297-332.

Kartiwa, Suwati

2000. *Shipcloths, Rare Treasures From Lampung*. Jakarta: Proyek Pembinaan Museum Nasional.

2007. *Ragam kain Tradisional Indonesia Tenun Ikat*. Jakarta: Gramedia.

Kartodirdjo, Sartono, Marwati D. Poesponegoro, Nugroho Notosusanto

1975. *Sejarah Nasional Indonesia (Jilid IV)*. Jakarta: Depatemen Pendidikan dan Kebudayan.

1977. *Sejarah Nasional Indonesia I-II*. Jakarta: Balai Pustaka.

Kartodirdjo, et al. (ed.)

1993. *700 Tahun Majapahit*. Surabaya: CV. Tiga Dara.

Kempers, A.J. Bernet

1959. *Ancient Indonesian Art*. Harvard University Press.

Kerlogue, Fiona

2000. 'Interpreting Textiles as a Medium of Communication: Cloth and Community in Malay Sumatra', *Asian Studies Review*, vol. 24, no. 3, September 2000.

Knapp, G & H. Sutherland.

2004. *Monsoon Traders: Ships, Skippers and Commodities in Eighteenth-Century Makassar*. Leiden: KITLV Press.

Knight-Achyadi, Judith & Asmoro Damais

2005. *Butterfloer and Phoenix - Chinese Inspirations in Indonesian Textiles Arts*. Jakarta: Mitra Museum Indonesia.

Koentjaraningrat

1989. *Pengantar Ilmu Antropologi*. Jakarta: Aksara Baru.

Kuhnt-Saptodewo, Sri (ed.)

2012. *Maluku: Sharing Cultural Memory*. Vienna: The Embassy of Republic Indonesia and Museum für Völkerkunde Wien.

Kusen

1981. *Arca-arca Terracotta Majapahit* (Sebuah Studi tentang Fungsi dan Kedudukannya). Karya Tulis Sarjana Muda UGM. Yogyakarta: FS-UGM.

Kusumaatmadja, M.

1978. *Bunga Rampai Hukum Laut*. Bandung: Binacipta.

Lapian, A.B.

1991. 'Sejarah Nusantara Sejarah Bahari', Pidato Pengukuhan Guru Besar Fakultas Sastra Universitas Indonesia. Jakarta

2009. *Orang Laut, Bajak Laut, Raja Laut: Sejarah Kawasan Laut Sulawesi Abad XIX*. Jakarta: Komunitas Bambu, 2009.

Leigh, Barbara

1989. *Tangan-Tangan Terampil: Seni Kerajinan Aceh*. Djambatan.

Leur, J.C. van

1955. 'Indonesian Trade and Society'. *Essays in Asiatic Society and Economic History*. The Hague: W. van Hoeve.

Li Zhiyan & Cheng Wen

1989. *Chinese Pottery and Porcelain: Traditional Chinese Arts and Culture*. Second printing. Beijing: Foreign Languages Press.

Lombard, Denys

2006. *Kerajaan Aceh Zaman Sultan Iskandar Muda (1607-1636)*. Jakarta: Kepustakaan Populer Gramedia.

Mac Donnel, Arthur Antony 1954. *A Practical Sanskrit Dictionary*. Oxford: University Press.

Macknight, C.C.

1976. *'The Voyage to Marege': Macassan Trepangers in Northern Australia*. Melbourne: Melbourne University Press.

Madjid, Nurcholis

1987. *Islam, Kemoderan dan Keindonesiaan*. Bandung: Mizan.

Marrison, G.E

1955. 'Persian Influences in Malay Life 1280-1650'. *JMBRAS*, vol. XVIIII, 1, pp. 52-69.

Marsden, William

2008. *Sejarah Sumatra*. Jakarta: Komunitas Bambu.

Martowikrido, Wahyono

1999. *Indonesian Gold: Treasures from the National Museum, Jakarta*. Brisbane: Queensland Art Gallery.

Maulana, Ratnaesih

1997. *Ikonografi Hindu*. Fakultas Sastra Universitas Indonesia.

Maziyah, Siti et al. (ed.)

2015. *Ornamen Mantingan: Koleksi Museum Jawa Tengah Ranggawarsita*. Semarang: Museum Jawa Tengah Ranggawarsita, Dinas Kebudayaan dan Pariwisata Jawa Tengah.

Moquette, J.P.

1913. 'De Eerste Vorsten van Samoedra-Pase (Noord-Sumatra)'. *R.O.D*, pp. 1-12.

Munandar, Agus Aris

2003. 'Arca Prajnaparamita sebagai Perwujudan Tokoh', *Aksamala*, Seri Kajian Arkeologi. Jakarta: Akademia.

Munawar, Tuti et al.

1986. *Koleksi Museum Nasional Jilid*, III. Jakarta: Proyek Pengembangan Museum Nasional.

Mundardjito et al.

2006. *Majapahit-Trowulan*. Jakarta: Direktorat Peninggalan Purbakala and Indonesian Heritage Society.

Netsher, E. & J.A .van der Chijs

1864. 'De Munten van Nederlandsch Indië: Beschreven en afgebeeld'. *VBG*, XXXI, p. 165.

Noordyn

1972. *Islamisasi Makassar*. Jakarta: Bhratara.

Nooteboom, Ch.R.

1932. *De Boomstamkano in Indonesië*. Leiden: Brill.

Paeni, Mukhlis (ed.)

2009. *Sejarah Kebudayaan Indonesia: Sistem Teknologi (hlm 153)*. Jakarta: PT Rajagrafindo Persada.

Pelly, U.

2013. *Ara dengan Perahu Bugisnya: Pinisi Nusantara*. Yogyakarta: Casa Mesa Publisher [1st edition 1975].

Perret, Daniel

2002. 'Batu Aceh: Empat Negara Asia Tenggara, Satu Kesenian'. *25 Tahun Kerjasama Pusat Penelitian Arkeologi dan École Française d'Extrême-Orient*. Jakarta.

Perret, Daniel & Heddy Surachman

2007. 'Jejak-Jejak Persia di Barus'. *Amerta*. Jakarta: Puslitbang Arkenas, pp. 1-11.

Peter, S. Dance

1992. *Shells*. Australia: Harper Collins Publisher.

Pijper, G.F.

1992. *Empat Penelitian Tentang Agama Islam di Indonesia 1930-1950*. Terj: Tudjumah. Jakarta: UI Press.

Poelinggomang, Edward L.
2002. *Makassar Abad XIX: Studi tentang Kebijakan Perdagangan Maritim*. Jakarta: Kepustakaan Populer Gramedia
Poelinggomang, Edward L., Bambang Budi Utomo & Trigangga (ed.)
2016. *The Maritime Legacy of Indonesia*. Jakarta: National Museum of Indonesia.
Poesponegoro, Marwati Djoened & Nugroho Notosusanto
1990. *Sejarah Nasional Indonesia II*. Depdikbud. Jakarta: Balai Pustaka.
Post, Peter
1991. *Japanese Bedrijvigheid in Indonesia, 1868-1942*. Centrale Huisdrukkerij Vrije Universiteit, Amsterdam.
Ramelan, Wiwin Djuwita S. (ed.)
2013. *Candi Indonesia Seri Jawa*. Jakarta: Direktorat Pelestarian Cagar Budaya dan Museum, Kementerian Pendidikan dan Kebudayaan.
2014. *Candi Indonesia Seri Sumatera, Kalimantan, Bali, Sumbawa* . Jakarta: Direktorat Pelestarian Cagar Budaya dan Museum, Kementerian Pendidikan dan Kebudayaan.
Rawson, Philip
1967. *The Art of Southeast Asia*. London: Thames and Hudson.
Reichel, Natasha
2007. *Violence and Serenity: Late Buddhist Sculpture from Indonesia*, Honolulu: University of Hawai'i Press.
Reid, Anthony
1988. *Southeast Asia in the Age of Commerce 1450–1680, vol. I: The Lands below the Winds*. New Haven.
1993. *Southeast Asia in the Age of Commerce 1450–1680, vol. II: Expansion and Crisis*. New Haven.
Ricklefs, M.C.
1981. *A History of Modern Indonesia since ca. 1300*. London: Macmillan.
Ridho, Abu
1979. *Oriental Ceramics, the World's Great Collection, Museum Pusat, Jakarta*, vol. 3. Japan: Kodansha Ltd.
1991. 'Han Dynasty Ceramics in Indonesia'. *Spirit of Han: Ceramics for the After-Life*. Southeast Asian Ceramic Society.
Riederer, Josef
1994. 'The Goldsmith's techniques: The technological analysis of early gold objects from Java', *Old Javanese Gold (4th-15th century): An archaeometrical approach*, Wilhelmina H. Kal (ed.). Amsterdam: Royal Tropical Institute.
Robert, Wessing
1997. 'Nyai Roro Kidul in Puger: Local Applications of a Myth'. *Archipel*, vol. 53, 1997, pp. 97-120.
Roelofsz, M.A.P. Meilink
1962. *Asian Trade and European Influence in The Indonesia Archipelago Between 1500 and about 1630*. The Hague: Martinus Nijhof.

Romuty, J.N.
1967. *Arti dan fungsi Pela di Ilwjar Wkmjer (Pulau-pulau Barbar)*. Skripsi. tidak diterbitkan. Institut Keguruan dan Pendidikan Djakarta-Cabang Ambon.
Rouffaer, G.P.
1980. *Cab Sikureueng (Segel Sultan Aceh)*. Daerah Istimewa Aceh: Proyek Rehabilitasi dan Perluasan Museum.
Santiko, Hariani
1987. 'Hubungan Seni dan Religi Khususnya dalam Agama Hindu di India dan Jawa', *Diskusi Ilmiah Arkeologi II: Estetika dalam Arkeologi Indonesia*. Jakarta: Pusat Penelitian Arkeologi Nasiona.
1992. *Bhatari Durga*. Fakultas Sastra Universitas Indonesia.
2003 'Perkembangan Awal Agama-agama di Indonesia'. In: [cat. exh.] *Fajar Masa Sejarah Nusantara*. Jakarta: Museum Nasional, pp. 41-48.
Satari, Soejatmi (ed.)
2009. [cat. exh.] *Treasures of Sumatra*. Jakarta: Museum Nasional.
Satari, Soejatmi et al.
1997. *Untaian Manik-Manik Nusantara*. Jakarta: Proyek Permuseuman.
Satria, Deddy
2005. 'Kaligrafi Arab: Khat Tsulust Ornamen Pada Seni Bangunan Makam di Aceh'. *Arabesk*. Nanggroe Aceh Darussalam: BP3 Banda Aceh, pp. 23-34.
Schoulten, C.
1953. *The Coin of The Dutch Overseas Territories 1601-1948*. Amsterdam: J. Schulman.
Sedyawati, Edi (ed.)
2012. *Indonesia dalam Arus Sejarah: Kerajaan Hindu-Buddha*. Jakarta: P.T. Ichtiar Baru van Hoeve dan Kementerian Pendidikan dan Kebudayaan Republik Indonesia.
Shimizu, Hiroshi
1988. 'Dutch-Japanese Competition in the Shipping Trade on the Java-Japan Route in the Inter-War Period'. In: *Tonan Ajia Kenkyu*, vol. 26, 1 (June), 1988, pp. 3-23.
Smen, Rudolf G.
2004. *Collection Batik from The Courts of Java and Sumatra*. Singapore: Periplus Edition.
Smidt, Dirk (ed.)
1993. *Asmat Art: Woodcarving of Southwest New Guinea*. Leiden: Periplus Editions & The Rijksmuseum voor Volkenkunde.
Soejono, R.P & R.Z. Leirissa (ed.)
2008. *Sejarah Nasional Indonesia II: Zaman Kuno*, Edisi Pemutakhiran. Jakarta: Balai Pustaka, 2nd ed.
Soejono, R.P. & H.R. van Heekeren
2008. *The Bronze-Iron Age of Indonesia*. The Hague: Martinus Nijhoff.

Soekmono, R.
1965. 'Archaeology and Indonesian History'. *An Introduction to Indonesian Historiography*. Soedjatmoko (ed.). Ithaca.
1973. *Pengantar Sejarah Kebudayaan Indonesia*. Jakarta: Kanisius.
1981. *Pengantar Sejarah Kebudayaan Indonesia 2* (3de ed.). Jakarta: Penerbitan Yayasan Kanisius.
Soemadio, Bambang et al.
1996/1997. *Buku Petunjuk Koleksi Arkeologi*. Proyek Pembinaan Museum Nasional Jakarta. Jakarta: Museum Nasional.
Soeroto
1976. *Sriwijaya Menguasai Lautan*. Bandung, Jakarta: Sanggabuwana.
Stenross, Kurt
2011. *Madurese Seafarers: Prahus, Timber and Illegality on the Margins of the Indonesian State*. Singapore: NUS Press.
Subarna, Abay D.
1999. 'Empat Gambar Untuk Sang Guru'. In: Denys Lombard, *Panggung Sejarah, Persembaha Kepada* (ed. Henry Chambert-Loir & Hasan Muarif Ambary). Jakarta: EFEO, Puslit Arkenas dan Yayasan Obor Indonesia.
Sudjarwo, Heru & Undung Wiyono
2010. *Rupa dan Karakter Wayang Purwa*. Jakarta: Kakilangit Kencana.
Sufi, Rusdi
1997. 'Mata Uang Kerajaan-Kerajaan di Aceh'. *Pasai Kota Pelabuhan Jalan Sutra: Kumpulan Makalah Diskusi*. Jakarta: Depdikbud.
Sugiyanti, Sri
1989. 'Ragam Hias Mesjid Mantingan, Jepara', *Proceedings Pertemuan Ilmiah Arkeologi V*, Yogyakarta, 4-7 July 1989. Jakarta: IAAI.
Suhardini, Sulaiman Yusuf *Aneka Ragam Hias Tenun Indonesia*.
Suleiman, Satyawati
1979. 'Penelitian Sejarah dan Sejarah Kesenian Sriwijaya'. In: *Pra Seminar Penelitian Sriwijaya Jakarta 7-8 December 1978*, Pusat Penelitian Arkeologi Nasional. Jakarta: PT Rora Karya.
1981 *Monuments of Ancient Indonesia* (second edition). Jakarta: Pusat Penelitian Arkeologi Nasional.
1981. *Sculptures of Ancient Sumatra*. Jakarta: Pusat Penelitian Arkeologi Nasional.
1983. 'Artinya Penemuan Baru Arca-arca Klasik di Sumatra untuk Penelitian Arkeologi Klasik'. *Rapat Evaluasi Hasil Penelitian Arkeologi I*. Jakarta: Pusat Penelitian Arkeologi Nasional, pp. 201-221.
Sulistioningsih, Retno & John N. Miksic (ed.)
2006. *Icons of Art: National Museum Jakarta*. Jakarta: BAB Publishing.
Sulistiyono, Singgih Tri
2003. *The Java Sea Network: Patterns in the Development of Interregional Shipping and Trade in the Process of Economic Integration in Indonesia, 1870s-1970s*. Dissertation Leiden University.

Sumadio, Bambang, Sulaiman Yusuf, Hamzuri & Dadang Udansyah
1980. *Koleksi Pilihan Museum Nasional Jilid I*. Jakarta: Proyek Pengembangan Museum Nasional.

Summerfield, Anne & John
Walk in Splendor: Ceremonial Dress and the Minangkabau. Los Angeles: UCLA Fowler Museum of Cultural History.

Sundari, Ekowati
2014. 'Fungsi dan Makna pada Beberapa Koleksi Keramik Museum Nasional'. *Prajnaparamita Jurnal Museum Nasional*, ed.: 02/2014. Jakarta: Museum Nasional, Ditjenbud,Kemendiknas.

Suropati, Untung, Yohanes Sulaiman & Ian Montana
2016. *Arungi Samudra Bersama Sang Naga: Sinergi Poros Maritim Dunia dan Jalur Sutra Maritim Abad ke-21*. Jakarta: PT Elex Media Komputindo - PT Gramedia.

Susetyo, Sukowati
2014. 'Hiasan Makara pada Masa Sriwijaya'. *Seminar Sriwijaya dalam Konteks Wilayah Asia Tenggara dan Asia Selatan*, Jambi: Pusat Arkeologi.

Supriyatun, Rini
1986. 'Arca Prajnaparamita dari Muara Jambi'. In: *Pertemuan Ilmiah Arkeologi*, IV, Cipanas 3-9 Maret 1986. Jakarta: Pusat Penelitian Arkeologi Nasional.

Sutrisno, M.M. (ed.)
2001. *Peta Indonesia Dari Masa ke Masa*. Jakarta: Museum Nasional.

Taylor, Paul Michael & Lorraine V.Aragon.
1991. *Beyond the Java Sea: Art of Indonesia's Outer Islands*. Washington/D.C: The National Museum of Natural History.

Tim Ekspedisi
1991. *Nelayan Lamalera*. Yogyakarta: Fakultas Sastra Universitas Gadjah Mada.

Tjandrasasmita, Uka
1975. *Sejarah Nasional Indonesia III: Jaman Pertumbuhan dan Perkembangan Kerajaan-kerajaan Islam di Indonesia*. Jakarta: Departemen Pendidikan dan Kebudayaan.
1993. 'Majapahit dan Kedatangan Islam serta Prosesnya'. *700 Tahun Majapahit (1293-1993) Suatu Bunga Rampai*. Dinas Pariwisata Daerah Propinsi Daerah Tingkat I Jawa Timur, Surabaya, pp. 277-289.
2000. 'Hubungan Perdagangan Indonesia-Persia Pada Masa Lampau dan Dampaknya Terhadap Beberapa Unsur Kebudayaan'. *Jurnal Pemikiran Islam Kontekstual IAIN Syarif Hidayattuloh*.
2000. *Pertumbuhan dan Perkembangan Kota-Kota Muslim di Indonesia*. Kudus: Menara Kudus.
2008. *Sejarah Nasional Indonesia III, Zaman Pertumbuhan dan Perkembangan Kerajaan Islam di Indonesia*; 2nd ed. Jakarta: Balai Pustaka.
2009. *Arkeologi Islam Nusantara*. Jakarta: KPG-EFEO-Fak. Adab & Humaniora UIN Syarif Hidayatullah.

Turner, Jack
2011. *Sejarah Rempah: Dari Erotisme hingga Imperialisme* (Penerjemah: Julia Absari). Jakarta: Komunitas Bambu.

Untoro, Heriyanti Ongkodharma
1999. 'Perdagangan Lada di Kesultanan Banten'. *Cerlang Budaya:Gelar Karya untuk Edi Sedyawati*. Depok: PPKB-LPUI.
2003. 'Menerapkan Arkeologi-Ekonomi Mengungkap Masa Lalu'. *Cakrawala Arkeologi: Persembahan untuk Prof. Dr. Mundardjito*. Depok: Jurusan Arkeologi-FIB, pp. 101-105.

Vainker, S.J.
1997. *Chinese Pottery And Porcelain: From Prehistory To The Present*. Reprinted, London: British Museum Press.

Van Der Hoop, N.J. Th. A Th.
1949. 'Indonesische Siermotieven'. Koninklijk Bataviaasch Genootschap van Kunsten en Wetenschappen.

Van Leur, J.C.
1955. *Indonesia Trade and Society: Essays in Asian Social and Economic History*. The Hague - Bandung: W. van Hoeve.

Vuuren, L. van
1917. 'De prauwvaart van Celebes', *Koloniale Studiën*, pp. 107-116; pp. 229-339.

Wardhani, Fifia
2015. 'Makara in Temple of Old Classical Era of Indonesia'. *ANGIS and CRMA Bangkok Meeting*. Bangkok: Princess Maha Chakri Sirindhorn Anthropology Centre, 5-6 January 2015.

Waren, J.F.
1981. *The Sulu Zone 1768-1898: The Dynamics of External Trade, Slavery, and Ethnicity in the Transformation of a Southeast Asian Maritime State*. Singapore: Singapore University Press.

Warsidi, A. et al.
2012. *Wisata Bahari Indonesia*. Jakarta: Tempo.

Waterson, Roxana
1991. *The Livingg House: An Anthropology of Architecture in South-East Asia*. Singapore: Oxford University Press.

Wheatley, P.
1980. *The Golden Khersonese*. Kuala Lumpur: Universiti Malaya Press.

Wiradnyana, Ketut
1999. 'Ekskavasi Situs Kerang Pangkalan Kabupaten Aceh Timur'. In: *Berkala Arkeologi 'Sangkhakala'*. Medan: Kementerian Kebudayaan dan Pariwisata.
Perahu, simbol perubahan makna pada tradisi Megalitik di Masyarakat Nias Selatan, https://www.academia.edu/3450045/ PERAHU_SIMBOL_PERUBAHAN_MAKNA_ PADA_MASYARAKAT_NIAS_SELATAN

Wirjosuparto, Sutjipto
1962. 'Sejarah Bangunan Mesjid di Indonesia'. *Almanak Muhammadiyah* Tahun 1381 H. No. XXI. Jakarta: Pimpinan Pusat Muhammadiyah, Majelis Taman Pusaka.

Wolter, O.W.
1967. *Early Indonesia Commerce: A Study of the Origin of Srivijaya*. Ithaca, NY: Cornell University Press.

Yahaya, Mahayudin Haji
1998. *Islam di Alam Melayu*. Kuala Lumpur: Dewan Bahasa dan Pustaka.

Yatim, Othman Mohd
1987. *Batu Aceh: Early Islamic Gravestones in Peninsular Malaysia*. Kuala Lumpur: Museum Association of Malaysia.

Yatim, Othman Mohd & Abdul Halim Nasir
1990. *Epigrafi Islam Terawal di Nusantara*. Kuala Lumpur: Dewan Kajian Bahasa, Kementrian Pendidikan Malaysia.

Zoetmoelder, P.J.
1983. *Kalangwan: Sastra Jawa Kuno Selayang Pandang*. Jakarta: Penerbit Djambatan.

Zuhdi, Susanto & Trigangga
2004. [cat. exh.] *Perlawanan dan Perjuangan Bangsa Indonesia Terhadap Kolonialisme*. Jakarta: Museum Nasional.

COLOPHON

Under the High Patronage of H.E. Mr. Joko Widodo, the President of the Republic of Indonesia and Their Majesties the King and Queen.

PROGRAMME & ORGANISATION OF EUROPALIA INDONESIA

IN INDONESIA

MINISTRY OF EDUCATION AND CULTURE REPUBLIC OF INDONESIA

DIRECTORATE GENERAL OF CULTURE

Director General
Hilmar Farid

Secretariat of Directorate General of Culture
Nono Adya Supriyanto, Secretary
Wawan Yogaswara
Sugeng Riadi
Fitra Arda
Kosasih Bismantara
Darmawati
Brimoresa Wahyu Dhoran Dhoro

Directorate of Cultural Property Preservation and Museums
Harry Widianto, Director
Judi Wahjudin
Ni Ketut Wardani Pradnya Dewi
Ridwan Muhammad Natsir Muslimin
Guntur

Directorate of Heritage and Cultural Diplomacy
Nadjamuddin Ramly, Director

Sub-Directorate Foreign Cultural Diplomacy
Ahmad Mahendra
Darwin Tampubolon
Ruliah Hasyim
Nur Lina Chusna
Gentur Adiutama
IGN Gde Dyaksa
Dina Amelia
Lilian Darra Bella

Sub-Directorate Programme, Evaluation, and Documentation
Roseri Rosdy Putri
Waluyo Agus Priyanto
Sinatriyo Danuhadiningrat

Directorate for the Arts
Restu Gunawan, Director

Sub-Directorate Visual Arts
Pustanto
Darmansyah
Joko Madsono
Setianingsih
Maulina Ratna Kustanti
Guntur Eka Budhi P.

Mardhiyas Citra Handriyani
Anita Noer Rachmi Hardini

Sub-Directorate Performing Arts
Yusmawati
Wahdat
Meta Ambarpana
Bunga Syamsu Wirandani
Galih Setyono

Sub-Directorate Media Art
Edi Irawan
Ibnu Sutowo
Suparman
Edi Suprianto
Rahman Ahkam

Sub-Directorate Programme and Evaluation
Kuat Prihatin
Sri Kuwati
Ike Rafiqah
Dindawati Fatimah
Agus Irawan

National Museum of Indonesia

National Gallery of Indonesia
Tubagus Sukmana, Director
Asikin Hasan, curator
Sumarmin
Firdaus

WORKING COMMITTEE

General Commissioner
EUROPALIA INDONESIA
Shanti L. Poesposoetjipto

Advisor
Abduh Aziz

Arts & Exhibitions
Amna S. Kusumo, Artistic Coordinator
Dian Ika Gesuri
Vicky Rosalina

Visual Arts Curators
Daud Tanudirjo
Linda Rooseline Octina, Assistant
Riksa Afiaty
Hikmat Darmawan
Ni Nyoman Nanda Putri Lestari, Assistant
Mohammad Cahyo Novianto
Ayos Purwoadji, Assistant curator
Alia Swastika
Venti Wijayanti & Irham Nur Anshari, Assistants
Danny Wicaksono
Stephanie Larassati, Assistant
National Museum of Indonesia
National Gallery of Indonesia

Performing Arts Curators
Sal Murgiyanto
Afrizal Malna

Music Curator
Ubiet Raseuki
Yasmina Zulkarnain, Assistant

Literature Curators
Melani Budianta
Manneke Budiman

Cinema Curator
Nan Achnas
Meninaputri Wismurti, Assistant

Communication
Yayoe Pribadi
Ratna Pandjaitan
Idham Setiadi
Dahlia Sardjono
Dede AM
Amalia Prabowo

IN BELGIUM
EUROPALIA ARTS FESTIVAL
WORKING COMMITTEE IN BELGIUM

General Commissioner
EUROPALIA INDONESIA
Michèle Sioen

General Management
Koen Clement
Kristine De Mulder (until 12 '16)

Arts
Dirk Vermaelen, Artistic director
Eva Bialek
Bozena Coignet
Marleen De Baets
Christoph Hammes
Marie-Eve Tesch
Lara Groeneweg, Intern
Rachel Katherina, Intern

Communication & Sponsoring
Colette Delmotte
Aurore Detournay-Kaas
Arnaud de Schaetzen
Axelle Desmaele
Maarten Huysmans
Thi Tuong Vi Vo
Inge De Keyser
Aurore Jeannot, Intern

Finance, Human Resources & General Administration
Stefana Ciubotariu, Director
Anne Doumbadze
Julie Erler
Van Ly Nguyen

Press
Performing Arts: Séverine Provost, Lies Gilis & Astrid Dubié BE CULTURE
Exhibitions: Gerrie Soetaert, Press & Communication

Performing Arts curator
Arco Renz

Music curator
Bart Barendregt

**BOARD OF DIRECTORS
EUROPALIA INTERNATIONAL**

Chairman
Count Georges Jacobs de Hagen

Vice-Chairman
Baron Jan Grauls

Members
Baron Paul De Keersmaeker, honorary
chairman
Viscount Etienne Davignon
Philippe Delaunois
Baron Jean Stephenne
Baron Rudi Thomaes
Baron Bernard Snoy
Count Paul Buysse
Regnier Haegelsteen
Véronique Paulus de Châtelet
Dirk Renard
Herman Daems
Freddy Neyts
Baron Pierre-Olivier Beckers
Baron Luc Bertrand
Baron Pierre Alain De Smedt
Alexis Brouhns
Baron Philippe Vlerick
Count Herman Van Rompuy
Christophe Convent

BNP Paribas Fortis Bank
Brusselse Hoofdstedelijke Regering –
Région de Bruxelles-Capitale
Council of Europe
Belfius Bank
National Lottery
FPS Foreign affairs
Fédération Wallonie-Bruxelles
Belgian Science Policy
Vlaamse gemeenschap
Deutschsprachige gemeinschaft
National Bank of Belgium
European parliament

Special thanks to:
The FPS Foreign Affairs
The Embassy of Belgium– Jakarta and
especially H.E. Ambassador and Madame
Hermann.

all our national and international partners
and artists, the numerous experts, con-
noisseurs and enthusiasts who helped us,
all the volunteers, all the subsidients and
sponsors who contributed to the success
of the festival

**ΞUROPALIA
ARTS FESTIVAL
INDONΞSIA**

EXHIBITION

**ARCHIPEL. KINGDOMS OF THE SEA
LIÈGE, LA BOVERIE,
25 OCTOBER 2017 – 21 JANUARY 2018**

CURATORS
Dra. Intan Mardiana Napitupulu
Les musées de la Ville de Liège

SCIENTIFIC ADVISORS
Prof. Dr. Singgih Tri Sulistiyono
Prof. Dr. Gusti Asnan
Abd. Rahman Hamid, Mhum
Drs. Bambang Budi Utomo, APU

ORGANISATION IN INDONESIA

Team Europalia Indonesia
Amna S. Kusumo, Artistic coordinator
Vicky Rosalina, Coordinator of exhibitions

*Directorate of Cultural Property Preservation
and Museums*

National Museum of Indonesia
Siswanto, Director
Intan Mardiana N., former Director

Dedah R. Sri Handari
Trigangga
Widodo
Dani Wigatna
Ita Yulita
Sri Suharni
Desrika Retno Widyastuti
Ni Luh Putu Chandra Dewi
Gunawan
Dyah Sulistyani
Karamina Puspitasari
Budiman
Valentina Beatrix Sondag
Fifia Wardhani
Ekowati Sundari
Dian Purwananta
Haryanti
Dwi Lestari
Muchlis Suharto

Lenders in Indonesia
National Museum of Indonesia, Jakarta
Museum of Lampung, Province Ruwa
Jurai
Museum of Siginjei, Jambi
The Cultural Heritage Preservation, Office
of Jambi (Balai Pelestarian Cagar Budaya
Jambi)
Museum Ranggawarsita, Semarang
Titarubi
Padewakang team (Bulukumba South-
Sulawesi)

Transport
DB Schenker

Inssurance
Lloyd's

Acknowledgments
Directorate for Cultural Properties and
Museums (Direktorat Pelestarian Cagar
Budaya dan Museum)
The National Research Center of
Archaeology (Pusat Penelitian Arkeologi
Nasional)
The National Library of the Republic of
Indonesia (Perpustakaan nasional
Republik Indonesia)
The National Archives of the Republic
Indonesia (Arsip Nasional Republik
Indonesia)
The Conservation Office of Borobudur
(Balai Konservasi Borobudur)
The Archaeological Office of South
Sumatera (Balai Arkeologi Sumatera
Selatan)
The Archaeological Office of Yogyakarta
(Balai Arkeologi Yogyakarta)
The Cultural Heritage Preservation Office
of Jambi (Balai Pelestarian Cagar Budaya
Jambi)
The Cultural Heritage Preservation Office
of Central Java (Balai Pelestarian Cagar
Budaya Jawa Tengah)
Museum of Central Java Province
'Ranggawarsita'
Museum of Lampung Province 'Ruwa Jurai'
Museum of Siginjei Jambi
Government of Bulukumba (South-Sulawesi)
Dian Soni Amellia; Drs. Bambang Budi
Utomo (photo); Fendi Siregar (photo-
video); Muchlis Soeharto (Photo) Budiman,
MA; Nanang Dwi Prasdi (video); the Team
of Museum Nasional Indonesia: Nusi
Lisabila E.; Rodina Satriana; Wahyu
Ernawati; Dian Novita; Imam Santoso;
Farah Ditha Hasanah; Kartina Risma
Wardhani; Katrynada Jauharatna; Nani
Mawarni; Dimas Seno Bismoko; Dita
Nurdayati; Ary Setyaningrum; Rahmita
Nurfitri Hapsari; Satria Putra Mandiri;
Muhammad Fahmi; Febriana Ika Saputri;
Dina Serepina; Pandu Satrio Atmojo
Ghautama; Ditta Nirmala; Hendrawan-
syah; Suswadi; Wiyarto; Suhanda; Adi
Suyitno; Sunarno; Muji S; Suroyo; Koma-
rudin; Sutikno; Carum; Wawan Agus;
Hamsidah; Siti Amanah; Nopa; Suharni;
Security team (Satuan Pengaman).
And the entire museum team.

ORGANISATION IN EUROPE

Working Committee
Fawzi Amri
Catherine Barsics
Pauline Bovy
Laura Dombret
Jean-Marc Gay
Jean-Marc Huygen
Audrey Jeghers
Christophe Remacle
Geoffrey Schoefs
Emmanuelle Sikivie
Dirk Vermaelen,
Marie-Eve Tesch

Scientific advisors
Prof. Pierre-Yves Manguin
(École française d'Extrême-Orient)
Prof. Catherine Noppe
(Musée Royal de Mariemont)
Alexis Sonet
(Musée Royal de Mariemont)
Bruno Hellendorff
(Groupe de recherche et d'information
sur la paix et la sécurité)

Lenders outside of Indonesia
Musée Royal de Mariemont
Musée de la Marine de Paris
Maritiem museum Rotterdam
Utrecht University
École française d'Extrême-Orient
Pierre-Yves Manguin
Musée de la Vie wallonne

Exhibition design
Jean-Marc Huygen

Graphic design
Jean-Marc Huygen
Karim Rezgui
Caroline Kleinermann
Maria Gallo

Exhibition production
Mosabois
Société Metrac
Les équipes techniques de la Ville de Liège

Conservation
The conservation department of the
Musées de Liège the National Museum
of Indonesia

Insurances
Ethias

Transport in Europe
Hizkia Van Kralingen

Audioguide
CloudGuide

Acknowledgments
Véronique Degroot, Horst Liebner, Denise
Biernaux and everybody who helped to
realise the exhibition and catalogue.

PARTNER IN BELGIUM

LA BOVERIE
BEAUX-ARTS · EXPO · LIÈGE

Exhibition organiser
asbl CIAC
Jean Pierre Hupkens, President
Pierre Gilissen, Michel Firket,
Vice-Presidents
and the Members of the Board of
Trustees and the General Assembly

Director of Operations
Jean-Marc Gay

Head of Exhibitions
Jean-Marc Gay

CATALOGUE

Scientific director
Dra. Intan Mardiana Napitupulu, M.Hum

Director of the publication
Prof. Dr. Singgih Tri Sulistiyono, MHum

General coordination
Jean-Marc Gay
Hans Devisscher

Authors of the essays
Dr. Andi F. Yahya
Abd. Rahman Hamid, MHum
Drs. Bambang Budi Utomo, APU
Desrika Retno Widyastuti, SS
Dra. Ekowati Sundari, MHum
Prof. Dr. phil Gusti Asnan
NLP Chandra Dewi, SS
Prof. Dr. Singgih Tri Sulistiyono, MHum
Drs. Trigangga
Dra. Wahyu Ernawati

Authors of the catalogue entries
Dimas Seno Bismoko, SHum
Fifia Wardhani, Shum
Haryanti, Spd
Karamina Puspitasari, SSos
Kartina Risma Wardhani, SHum
Katrynada Jauharatna, SHum
Nani Mawarni, SS
Valentina Beatrix Sondag, SSos

Translation
Duncan Brown
Timothy Straud

Copy editing
Dr. Andi F. Yahya
Duncan Brown

Design
Griet Van Haute

Photography
Arkadius Ganden

Lithography
Steurs, Antwerp

Production and publishing
Snoeck Publishers, Ghent
www.snoeckpublishers.be

Printing and binding
Printer Trento, Trento (It.)

ISBN: 978-94-6161-429-2
Legal deposit: D/2017/0012/78

Photo credits
© Museum Nasional Indonesia -
photo Arkadius

© 2017, Europalia International; La Boverie
(asbl CIAC/Ville de Liège); Snoeck Publishers,
Ghent; the authors